THE BOUNDING APPROACH TO VLSI CIRCUIT SIMULATION

THE KLUWER INTERNATIONAL SERIES IN ENGINEERING AND COMPUTER SCIENCE

VLSI, COMPUTER ARCHITECTURE AND DIGITAL SIGNAL PROCESSING

Consulting Editor

Jonathan Allen

Other books in the series:

Logic Minimization Algorithms for VLSI Synthesis. R.K. Brayton, G.D. Hachtel, C.T. McMullen, and A.L. Sangiovanni-Vincentelli. ISBN 0-89838-164-9.

Adaptive Filters: Structures, Algorithms, and Applications. M.L. Honig and D.G. Messerschmitt. ISBN 0-89838-163-0.

Computer-Aided Design and VLSI Device Development. K.M. Cham, S.-Y. Oh, D. Chin and J.L. Moll. ISBN 0-89838-204-1.

Introduction to VLSI Silicon Devices: Physics, Technology and Characterization. B. El-Kareh and R.J. Bombard. ISBN 0-89838-210-6.

Latchup in CMOS Technology: The Problem and Its Cure. R.R. Troutman. ISBN 0-89838-215-7.

Digital CMOS Circuit Design. M. Annaratone. ISBN 0-89838-224-6.

THE BOUNDING APPROACH TO VLSI CIRCUIT SIMULATION

by

Charles A. Zukowski
Columbia University
New York

SPRINGER SCIENCE+BUSINESS MEDIA, LLC

Library of Congress Cataloging-in-Publication Data

Zukowski, Charles A.
The bounding approach to VLSI circuit simulation.

(The Kluwer international series in engineering and computer science. VLSI, computer architecture, and digital signal processing)
Bibliography: p.
Includes index.
1. Integrated circuits—Very large scale integration—Mathematical models. 2. Integrated circuits—Very large scale integration—Data processing.
I. Title. II. Series.
TK7874.Z85 1986 621.395 86-10925
ISBN 978-1-4684-9893-6 ISBN 978-1-4684-9891-2 (eBook)
DOI 10.1007/978-1-4684-9891-2

Originally published by Kluwer Academic Publishers in **1986**
Softcover reprint of the hardcover 1st edition **1986**

Contents

List of Figures ix

Preface xiii

1. INTRODUCTION 1

2. VLSI CIRCUIT SIMULATION 5
2.1 General Circuit Simulators 5
2.2 Digital MOS Circuit Simulators 7
2.3 Trading Accuracy for Speed 10
2.3.1 Approximating Simulators 10
2.3.2 Bounding Approaches 12

3. SIMULATION WITH BOUNDS 15
3.1 Bound Definition 16
3.1.1 Bounding General Subsets 18
3.1.2 Bounding the Excitations 21
3.1.3 Bounding the Models 25
3.1.4 Bounding the Behaviors 26
3.1.5 Forms of Bounds 28
3.2 Basic Strategy 30
3.2.1 Bounding the Behavior of Simple Circuits 31
3.2.2 Bounding the Behavior of Monotonic Subcircuits 32
3.2.3 Relaxation of Bounds 35
3.3 Simulating Various Digital MOS Subcircuits 41
3.3.1 Standard Restoring Logic 41
3.3.2 Pass Logic 43
3.3.3 Latches 44
3.3.4 Ring Oscillators 46
3.3.5 Schmitt Triggers 47
3.3.6 Bootstrap Drivers 49

3.4 Applications for Bounding Algorithms 50
3.4.1 Uncertainty Management 50
3.4.2 Worst Case Analysis 53
3.4.3 Input-Independent Analysis 55
3.5 Summary 58

4. THE VLSI CIRCUIT MODEL **59**
4.1 Elements in the VLSI Circuit Model 60
4.1.1 Resistors 60
4.1.2 Transistors 62
4.1.3 Capacitors 65
4.1.4 Element Composition Rules 67
4.2 The Resistor Subnetwork 70
4.2.1 Terminal Contraints 70
4.2.2 Internal Elements 74
4.2.3 One-Port Groups 77
4.2.4 Resistor Subnetwork Summary 82
4.3 The Transistor Subnetwork 84
4.3.1 Comparison with the Resistor Subnetwork 84
4.3.2 Series and Parallel Transistor Compositions 86
4.3.3 Terminal Constraints and Internal Elements 90
4.3.4 Transistor Polarity 94
4.3.5 Transistor Subnetwork Summary 94
4.4 The Capacitor Subnetwork 97
4.4.1 Evaluation of Circuit Dynamics 97
4.4.2 Waveform Relaxation and the Capacitor Subnetwork 100
4.4.3 Terminal Constraints 107
4.4.4 Internal Elements 109
4.4.5 Capacitor Subnetwork Summary 116
4.5 Node Circuits 117
4.5.1 Single Node Equation 118
4.5.2 Differential Equation Inequalities 119
4.5.3 Monotonicity of the Node Equation 121
4.5.4 Analysis of the Node Equation 124
4.6 Cluster Circuits 125
4.6.1 General Cluster Circuit 126
4.6.2 One-Way Restoring Logic Gate 130
4.6.2.1 Resistive Logic Gate Models 130
4.6.2.2 Interconnect Models 132
4.6.2.3 One-Way Combinational Logic Models 134
4.7 Summary 137

5. BOUND RELAXATION **139**
5.1 Theory 140
5.1.1 Definitions 141
5.1.2 Using Bounding Functions to Tighten Bounds 145
5.1.3 Using Bounding Functions to Generate Bounds 146
5.2 D.C. Analysis 149
5.2.1 Basic Strategy 149
5.2.2 Resistor Network Example 150
5.2.3 MOS Feedback Circuit Example 152
5.3 Transient Analysis 154
5.4 Resistive Mesh Circuits 161
5.4.1 Notation and Definitions 163
5.4.2 Bound Relaxation for RC Mesh Circuits 166
5.4.3 Linear RC Mesh Example 170
5.5 Summary 171

6. ALGORITHMS AND EXPERIMENTAL RESULTS **173**
6.1 One-Way Logic Gate Models 174
6.1.1 General Logic Gates 175
6.1.2 General Inverters 179
6.1.3 Inverters with Capacitive Loads 181
6.1.4 Experiments 183
6.2 Linear RC Mesh Circuits 186
6.2.1 Bounding Algorithm 187
6.2.2 Experiments 189
6.3 General Circuit Models 194
6.3.1 Partitioning and Scheduling 194
6.3.2 Device Models 196
6.3.3 D.C. Analysis 197
6.3.4 Transient Analysis 198
6.3.5 Experiments 199
6.4 Summary 207

7. CONCLUSION **209**
7.1 Future Work 209
7.2 Closing Comments 211

References **212**

Index **219**

List of Figures

Figure 2-1: A digital MOS circuit can be partitioned into blocks. 9
Figure 2-2: Two orthogonal approaches to the speed-accuracy tradeoff. 11
Figure 3-1: Exact and bounding simulators involve mappings between sets. 17
Figure 3-2: Sets can be bounded using intervals based on a partial ordering. 19
Figure 3-3: The behavior subset is interpreted through a specification function. 20
Figure 3-4: Ignoring correlations over time can produce unattainable bounds. 25
Figure 3-5: Nonmonotonic bounds can be simplified into monotonic ones. 29
Figure 3-6: Logic simulation waveforms can be represented with bounds. 30
Figure 3-7: Simplified circuits are analyzed inside a bounding algorithm. 31
Figure 3-8: Large MOS circuits can be subdivided into monotonic subcircuits. 33
Figure 3-9: Simple example of a coupled system. 36
Figure 3-10: Exact relaxation towards the solution for $\alpha = -0.5$ and $\alpha = +0.5$. 37
Figure 3-11: Relaxation of bounds for $\alpha = -0.5$ and $\alpha = +0.5$. 38
Figure 3-12: Bounding algorithms can analyze restoring logic gates efficiently. 42
Figure 3-13: Pass networks can contain dynamic storage nodes. 43
Figure 3-14: The behavior of latches can be tightly bounded. 45
Figure 3-15: Ring oscillators restore signals in voltage but not in time. 47
Figure 3-16: Schmitt triggers exhibit positive feedback. 48
Figure 3-17: Bootstrap drivers are sensitive to correlations. 49
Figure 3-18: Simple circuit model to illustrate uncertainty management. 51
Figure 3-19: Simple example of a circuit model containing uncertainty. 54
Figure 3-20: Bounds shrink the search space in input-independent analysis. 57

Figure 4-1: Nonlinear resistors can be bounded with linear ones. 61
Figure 4-2: *N*-elements and *P*-elements are four terminal resistive elements. 63
Figure 4-3: A physical transistor is modeled using external capacitor elements. 64
Figure 4-4: Incremental bounds on capacitors limit incremental capacitance. 66
Figure 4-5: The contribution of each element type is analyzed separately. 68
Figure 4-6: A CMOS NAND gate partitioned into four element subnetworks. 69
Figure 4-7: The resistor subnetwork $\mathcal{N}_R$. 71
Figure 4-8: The terminal constraints of $\mathcal{N}_R$ are monotonic. 73
Figure 4-9: Example of $\mathcal{N}_R$ with complex dependence on resistor G_2. 75
Figure 4-10: All internal nodes of $\mathcal{N}_R$ are inside one-port groups. 76
Figure 4-11: A simple bridge circuit that is nonmonotonic with respect to R_0. 78
Figure 4-12: R_p contains both linear and nonlinear resistors. 79
Figure 4-13: The operating point lies on the darkened portion of the curve. 80
Figure 4-14: Branches b_0, R_1, and R_2 form a series-parallel subgroup. 81
Figure 4-15: The n-channel transistor subnetwork $\mathcal{N}_N$. 85
Figure 4-16: General *N*-elements model NMOS pullups and transistor groups. 87
Figure 4-17: Series and parallel connections of general *N*-elements. 90
Figure 4-18: The terminal constraints of $\mathcal{N}_N$ are not necessarily monotonic. 91
Figure 4-19: A modified circuit can be used to calculate bounds. 92
Figure 4-20: The capacitor subnetwork $\mathcal{N}_C$ 98
Figure 4-21: Dataflow diagrams for evaluation of node equations. 100
Figure 4-22: Series combination of capacitors. 105
Figure 4-23: A number of terminal voltages can be solved at once. 110
Figure 4-24: Bounding the effect of internal elements in $\mathcal{N}_C$. 112
Figure 4-25: Circuit transformation that can simplify bounding of $\mathcal{N}_C$. 113
Figure 4-26: Circuit corresponding to single node equation. 118
Figure 4-27: Circuit corresponding to a typical cluster. 125

Figure 4-28: Resistive elements between cluster nodes can be transformed. 128
Figure 4-29: The transistor portions of MOS logic gates are *G*-elements. 131
Figure 4-30: Gate interconnect wires are modeled with an RC tree. 134
Figure 4-31: Each logic gate is modeled as a simple *L*-element. 135
Figure 4-32: Effect of reconvergent paths in an *L*-element. 136
Figure 5-1: For any interval *W*, *F* maps any element in *W* to one in *G*(*W*). 143
Figure 5-2: The bounding function *G* is applied repeatedly in theorem 5-2. 147
Figure 5-3: Bound relaxation can incorporate element uncertainty. 151
Figure 5-4: Bounds on the d.c. solution of a latch are desired. 153
Figure 5-5: The transient analysis of complex VLSI circuits can be partitioned. 156
Figure 5-6: The model used for an MOS inverter. 156
Figure 5-7: G_x produces bounds on the response of each inverter. 157
Figure 5-8: Partial table of currents for hypothetical CMOS inverter i, i=1,2,3. 158
Figure 5-9: Behavior intervals W^0 and $G_x(W^0)$ for the circuit of figure 5-5. 159
Figure 5-10: Linear RC mesh circuits are often used to model MOS circuits. 162
Figure 5-11: A typical node I of the linear RC mesh circuit 𝔅. 164
Figure 5-12: Equivalent circuit for node I of network 𝔅. 165
Figure 5-13: Example illustrating bound relaxation for linear RC mesh circuits. 170
Figure 6-1: Lower parallel bounds and upper series bounds (n-channel). 177
Figure 6-2: Bounding a transistor pair with a single transistor. 178
Figure 6-3: Linear lower bounds on a *G*-element output for a staircase input. 179
Figure 6-4: Linear RC tree that bounds interconnect response. 180
Figure 6-5: Using the response to a "typical" input waveform. 182
Figure 6-6: Integrating a staircase input. 184
Figure 6-7: Lower bounds on the output of a chain of 4 NAND gates. 186
Figure 6-8: APL global variables and programs used in experiments. 190
Figure 6-9: The RC line considered as the first example. 191

Figure 6-10: Three relaxations of L^0 and U^0. 191
Figure 6-11: Three relaxations of $\hat{L}$ and $\hat{U}$. 192
Figure 6-12: The bus driver considered as the second example. 193
Figure 6-13: Relaxations of L^0 and U^0 for nodes 1 and 3. 193
Figure 6-14: A chain of inverters with Miller capacitors. 200
Figure 6-15: Three relaxations for an inverter chain. 201
Figure 6-16: An NMOS latch. 202
Figure 6-17: Behavior bounds for latch with three pulse widths. 202
Figure 6-18: A CMOS NAND gate. 203
Figure 6-19: Behavior bounds for CMOS NAND gate. 204
Figure 6-20: An NMOS pass network. 205
Figure 6-21: Behavior bounds for an NMOS pass network. 205
Figure 6-22: An XOR gate with feedback. 206

PREFACE

This book proposes a new approach to circuit simulation that is still in its infancy. The reason for publishing this work as a monograph at this time is to quickly distribute these ideas to the research community for further study. The book is based on a doctoral dissertation undertaken at MIT between 1982 and 1985. In 1982 the author joined a research group that was applying bounding techniques to simple VLSI timing analysis models. The conviction that bounding analysis could also be successfully applied to sophisticated digital MOS circuit models led to the research presented here.

Acknowledgments

The author would like to acknowledge many helpful discussions and much support from his research group at MIT, including Lance Glasser, John Wyatt, Jr., and Paul Penfield, Jr. Many others have also contributed to this work in some way, including Albert Ruehli, Mark Horowitz, Rich Zippel, Chris Terman, Jacob White, Mark Matson, Bob Armstrong, Steve McCormick, Cyrus Bamji, John Wroclawski, Omar Wing, Gary Dare, Paul Bassett, and Rick LaMaire. The author would like to give special thanks to his wife, Deborra, for her support and many contributions to the presentation of this research. The author would also like to thank his parents for their encouragement, and IBM for its financial support of this project through a graduate fellowship.

THE BOUNDING APPROACH TO VLSI CIRCUIT SIMULATION

1. INTRODUCTION

The VLSI revolution of the 1970's has created a need for new circuit analysis techniques. Classical electrical simulation methods are no longer feasible for many large circuits being designed. The field of VLSI circuit simulation has arisen from the need to efficiently analyze large MOS circuits that are primarily digital in nature. Work in the field has concentrated on deriving specialized algorithms that efficiently exploit the special properties of digital MOS circuits, as well as deriving approximation methods to sacrifice accuracy for computational speed. Progress has been made, but designers' needs continue to outpace the abilities of simulation tools.

This book explores a complete bounding approach to VLSI circuit simulation. A bounding approach trades accuracy for speed, producing an estimate of circuit behavior for which uncertainty is bounded. A method to generate efficient, rigorous bounds on the response of a complex digital MOS integrated circuit model is proposed. The method builds upon the latest work in the field of VLSI circuit simulation, and provides a framework in which further results can be easily incorporated and evaluated.

Substantial progress has recently been made in the development of specialized algorithms for the exact simulation of large digital MOS circuits. Many of these algorithms use some form of partitioning both to exploit latency and to achieve a computation time that scales linearly with circuit size. New algorithms such as Waveform Relaxation enable the exact simulation of circuits with thousands of transistors. These specialized algorithms are discussed, along with a review of the entire field of VLSI circuit simulation, in the second chapter. These new algorithms offer an improvement over those used in classical simulators, but they are still inadequate for VLSI circuits that can contain millions of transistors.

Simulators have been developed that can analyze very large circuits, but they are forced to use either rough numerical algorithms or very simple circuit models that sacrifice accuracy. Linear and piecewise-linear device models have been used to create very fast algorithms. These approximation algorithms are also reviewed in the second chapter. The basic idea of injecting uncertainty into simulation to improve speed looks appealing since a large amount of uncertainty can often be tolerated, especially in noncritical paths through combinational logic circuits. Rough simulators have been useful for first pass design and timing verification, but they cannot manage uncertainty effectively because they do not measure it. Hence, many designers continue to seek the reassurance of an "exact" simulator at great cost.

The desire to measure uncertainty has spawned efforts to generate rigorous bounds on the behavior of some circuits. Upper and lower bounds on a solution can be used to produce both an estimate for the solution and a measure of worst case uncertainty. The second chapter concludes with a discussion of previous bounding algorithms. Bounding techniques have been successfully applied to very simple circuit models, such as RC tree circuits, but efficient algorithms for more complex models have remained beyond reach. If only the response of very rough circuit models can be bounded, most of the uncertainty present in the result is not measured. In this case rigorous bounds are only useful for guiding the choice of an approximation for the response of the rough model.

The potential advantages of a complete bounding approach are large. The primary advantage is the ability it provides to manage uncertainty effectively. If uncertainty can be measured, only the level of accuracy that is required in any part of a simulation need be generated. With an approximating approach, the level of accuracy does not generally vary throughout a circuit. Also, as circuits become larger and the number of clock cycles considered grows, the amount of accuracy needed to achieve the same level of confidence grows due to the statistics of approximate analysis. The application of uncertainty management is discussed further in the third chapter. Bounding algorithms have the nice property that they never give an incorrect answer. In the worst case, a bounding algorithm might not provide sufficient accuracy in a given subcircuit at an acceptable cost, forcing the use of more conventional exact algorithms for that subcircuit.

In addition to measuring the uncertainty arising from simplified circuit analysis, a bounding algorithm can potentially incorporate uncertainty that is inherent in any simulation, e.g., fabrication process variations and incompletely specified circuit inputs. As shown in the third chapter, tolerances on device characteristics and circuit inputs can be incorporated rigorously into bounds on circuit behavior.

Bounding approaches for general problems are often very difficult and of limited value. The field of interval arithmetic, which has often addressed very general problems, has developed a poor reputation. The simulation of digital MOS circuits represents a very special problem that, when approached with an "interval arithmetic" at the level of logic waveforms, does not exhibit many of the difficulties of general problems. The third chapter contains a discussion of how problems found in general bounding algorithms can be avoided in the domain of digital MOS circuit simulation. An overview of the basic bounding strategy proposed in this book is also presented there, and its performance on various common circuit forms is considered in general terms.

To prove rigorous properties of a mathematical circuit model, a well defined class of networks must be considered. Chapter four defines such a class, and discusses the necessity for each constraint placed on the model. The constraints are minimized to include most reasonable MOS circuit models while capturing their essential properties. The results presented in this book apply to a very general class of MOS circuit models, including arbitrary topologies and allowing many complex device models currently in use.

In addition to being generally feasible for digital MOS circuits, bounding algorithms can use modifications of existing efficient approximation algorithms. Due to special monotonic properties of the MOS circuit model derived in the fourth chapter, simple linear or piecewise-linear circuit models can be generated whose behaviors bound that of the original. By using algorithms developed to analyze these simple circuit models, rigorous upper and lower bounds on signal waveforms can be generated, producing approximations with bounded uncertainty. Uncertainty in the circuit inputs and device models can also be naturally incorporated in this strategy. The fourth chapter discusses the use of simplified models to generate bounds on the behavior of key classes of subcircuits that correspond to typical partitions for digital MOS circuits. Small clusters that correspond roughly to logic gates are considered, along with important special cases.

When considering a digital MOS circuit model with feedback, including local feedback produced by Miller capacitance, even simplified bounding models become quite complex. To generate tight bounds, simplified bounding models must also contain the feedback paths present in the original circuit. The techniques developed for efficient exact analysis of such circuits can also be extended to include bounds. More specifically, the efficient Waveform Relaxation algorithm can be extended to include waveform intervals. The behavior of small tightly coupled subcircuits can be bounded separately, and these bounds then used iteratively to improve global bounds. This extension, an integral part of the bounding strategy proposed in this book, is presented in the fifth chapter. The partitioning essential for efficient VLSI simulation can still be achieved in the bounding context. The performance of the algorithm can be roughly maintained, and bounds can be generated for even sophisticated circuit models with a computation time that scales roughly linearly with circuit size. The result can be a telescoping sequence of rigorous bounds, where the number of iterations provides an additional mechanism to trade computation for guaranteed accuracy.

The relaxation scheme provides a natural strategy for a general bounding simulator. A large portion of a typical MOS circuit model contains only a few small and standard blocks that can each be analyzed in detail independently. These blocks, which are well understood, are then analyzed sequentially to improve a current bound on the behavior of the entire circuit. A bounding simulator based on relaxation can be improved by further work on bounding the responses of simple and common blocks. The relaxation approach also potentially allows different strategies to be used in different portions of a circuit, ranging from rough timing analysis of noncritical digital blocks to exact simulation of analog blocks.

Simple experiments to ascertain the feasibility of the high-level bounding strategy proposed in this book are presented in the sixth chapter. The algorithms used serve as an example of how a detailed algorithm can be built with the theoretical results contained in the fourth and fifth chapters. Simple algorithms for linear RC models, intermediate-complexity gate array models, and more complex general models are discussed, and their performance on a variety of circuits is explored. The results are encouraging and suggest promising areas for future research.

As circuits become larger and more sophisticated bounding algorithms are developed, bounding techniques can potentially provide a very useful enhancement for VLSI simulation programs. The ultimate goal of a bounding approach is to produce a low-cost and reliable simulator for large digital MOS circuits. The ideas and results presented in this book constitute a framework for the development of bounding algorithms for this purpose.

2. VLSI CIRCUIT SIMULATION

The field of VLSI circuit simulation addresses the problem of the inadequacy of standard circuit simulation programs to handle the large circuits that are now being designed. This chapter begins with a brief discussion of standard circuit simulators, and then proceeds to cover the two main approaches that have been taken to simulate VLSI circuits. The second section considers simulators that have been tailored to exploit the properties of large digital MOS networks, the major components of today's VLSI circuits. A discussion of Waveform Relaxation, a specialized algorithm exploited later in this book, is included in this section. The third section considers algorithms that trade substantial accuracy to obtain increased speed. Rough timing simulators have been developed that use very simplified circuit models to inject unmeasured uncertainty, and bounds on the responses of very simple circuit models have been used to inject measured uncertainty.

2.1 General Circuit Simulators

General circuit simulators were developed shortly after large computers became widely available. Their goal was to be able to analyze most electronic circuits that could arise. The most well known of these programs is SPICE [1]. ASTAP is a similar program that has been widely used [2]. Models have been added to these programs to characterize most modern devices and they have been very successful. They have been used on a wide variety of circuits and have built up a high level of confidence in their users. Sometimes they fail to converge to the solution without some prodding by the user, but they rarely provide a poor estimate of how a given circuit model will behave.

General simulators must use fairly general numerical integration methods and as a result, require a large amount of computation. The large circuits often considered today have outgrown the capabilities of these programs. The field of VLSI circuit simulation has produced many alternatives for large digital MOS circuits, but SPICE is still widely used, even in VLSI design. Designers push it to its limits and consider only critical paths, due to the confidence they have in the accuracy of its results. Such a general simulator will always be useful for designing small cells, evaluating device models, and verifying new simulation tools [3].

In a general simulator, even formation of the network equations is a difficult task due to the wide variety of models that must be considered. SPICE uses a technique called Modified Nodal Analysis that, roughly speaking, uses Kirchhoff's current law (KCL) at each node, along with voltage controlled representations of the element constraints, to produce an equation for each node voltage in terms of the others. ASTAP, on the other hand, uses a technique called Sparse Tableau. The ASTAP method, where each element and topology constraint is considered separately, produces a system of equations that is much larger, but generates matrices with greater sparsity.

General simulators partition circuit analysis into individual time steps. The incremental construction of the solution reflects the progression through time that governs general physical systems. A uniform step size, chosen to produce the required accuracy in the most sensitive portion of the circuit, is used throughout due to the arbitrary coupling allowed between circuit elements. At each step, the next values for the circuit variables are estimated based on the circuit model and previously calculated values. One of a number of standard implicit numerical integration algorithms is used for this purpose, with careful monitoring of stability and truncation error.

When implicit methods are used to calculate the values of the circuit variables after a time step, a system of nonlinear algebraic equations is produced. The solution of this system generates the state of the circuit at the next time step. By using variations of Newton's method, the system of equations is linearized to generate a sequence of intermediate guesses that usually converges to the solution. Since connectivity is small for most circuits, sparse matrix techniques are often used to solve the linear system of equations in the inner loop.

Commercial general simulators are packaged with well defined user interfaces. Special techniques are often added for tasks such as d.c. and frequency domain analysis of circuits. The user has access to accuracy parameters and can specify a wide range of simulation modes. Many users have added interfaces between their circuit data bases and a general simulator for easy access. Circuits with more than about 100 devices, though, strain most systems due to the fact that computation requirements are large, and grow more quickly than linearly with the size of the circuit.

2.2 Digital MOS Circuit Simulators

By restricting the class of circuits considered, a simulator can realize significant savings in computation with specialized algorithms. Most of today's large circuits are primarily digital MOS, providing many properties that can be readily exploited in a simulator. While digital MOS simulators are not as robust and general as would be preferred, constraints on computation often make them the best feasible approach. There are many such programs that have been developed, exploring many possible approaches, but no "standard" program has been established. Examples include MOTIS [4] [5], SPLICE [6] [7], DIANA [8] [9], MACRO [10], and RELAX [11] [12] [13].

One simplification that is often used in digital MOS simulation is in the representation of the network equations. Since large digital circuits are often specified hierarchically and contain many repetitions of basic cells, the equations can be expressed with similar efficiency. In addition, the exclusive use of a few device types allows the use of fine-tuned table-lookup device models for speed in function evaluation. Special representations and designer flags, though, if required, can be a burden if a circuit is being extracted from a source with limited information such as a set of mask specifications.

Another simplification that is generally used in digital MOS simulation is partitioned analysis. Digital circuits contain mostly small blocks that are well understood and connected in a well defined, sparse manner. Coupling between blocks is often primarily in one direction. As a result, one-way macromodels are sometimes used for standard blocks. Network equations can be solved hierarchically in a manner that reflects the structure of the circuit. This technique is called multi-level simulation. Connectivity between blocks, i.e., fan-in and fan-out, does not grow significantly with circuit size in digital circuits, so partitioning can produce a simulation algorithm whose time increases roughly linearly with the size of the circuit. Again, if the partitioning cannot be automated, it can cause great inconvenience.

Yet another simplification that must be used in digital MOS simulation to obtain efficiency concerns latency. At any instant in time, most of the circuit variables in a digital circuit are not changing. Activity propagates through the circuit in ripples. Efficiency demands that large amounts of computation only be used for circuit blocks during the times that they are active. Partitioned analysis of the circuit allows latency to be exploited in a straightforward way, only updating active blocks at each step or using variable time steps among blocks. Event driven algorithms are often used to analyze simple one-way models, starting analysis of a block only after its inputs start changing and ending analysis when it has settled.

A final simplification often used in digital MOS simulators is to slightly ease accuracy requirements. Digital circuits do not tend to be very sensitive to small errors in circuit variables as the circuits consist mostly of restoring logic. Errors are erased over time much as noise is in actual circuit operation. Simple integration techniques can be used that maintain acceptable accuracy for a wide range of circuits. In addition, restrictions can be placed on the models that have a small effect on their accuracy[1]. When small approximations are used the simulator is often called a timing simulator.

When a digital MOS circuit is partitioned for analysis, the opportunity arises to use different types of analysis in different portions. This technique, called mixed-mode simulation, is illustrated in the program SPLICE. A designer can specify whether a portion of the circuit should be analyzed with high accuracy simulation, timing simulation, or simply logic simulation.

A new algorithm that has recently generated interest in the VLSI simulation community is Waveform Relaxation [11], used in the program RELAX. This algorithm is presented here as an example of digital MOS simulation, and it is presented in some detail because the simulation methods derived in this book build upon it. In the Waveform Relaxation algorithm, a circuit is partitioned into strongly connected blocks, each of which is analyzed independently over entire time intervals. The algorithm provides a good example of one that exploits the properties of digital MOS circuits mentioned earlier to produce an efficient simulator.

The Waveform Relaxation algorithm requires a circuit that is partitioned into strongly connected blocks. While these blocks could be specified as an input, the process can be automated due to the flexibility allowed in the partitioning. When reasonable restrictions are placed on the circuit model, such as requiring the existence of a positive capacitor between each node and ground, poor partitions have the effect of increasing computation, but they do not prevent the algorithm from converging to the solution. If a circuit model is cut at the gate of

---

[1] The disallowing of internodal coupling capacitance is an extreme example. In this case the effect of Miller capacitance is modeled with grounded input and load capacitors.

each transistor, what remains is a number of transistor clusters that are only connected to each other through the power supply and internodal coupling capacitors as in figure 2-1. Such clusters are a reasonable choice for the strongly connected blocks in combinational logic circuits.

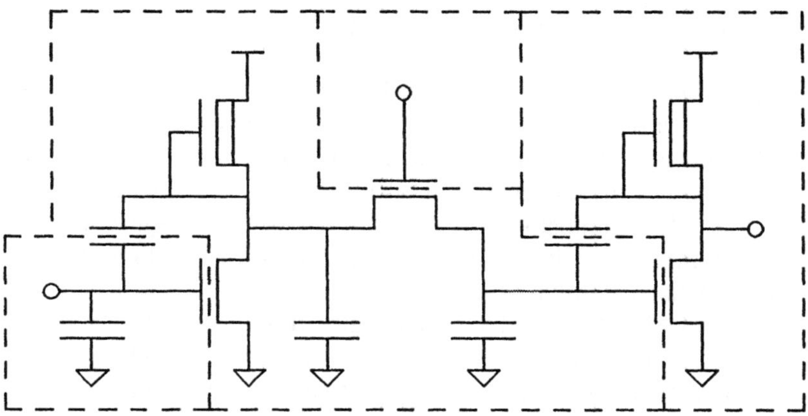

Figure 2-1: A digital MOS circuit can be partitioned into blocks.

Each of the blocks in the circuit is analyzed separately with numerical methods typical of more general simulators. The algorithm starts with any initial guess for the behavior of the circuit. The guesses for each block are then sequentially updated, using the latest guess for the behavior of all other blocks to compute all coupling among blocks. Latency is exploited because the time step used in the numerical simulation of each block is independent of that used for any other block. Such a procedure converges quickly for digital MOS circuits because coupling between the clusters is primarily in one direction, i.e., local feedback is weak. Internodal coupling capacitance is usually relatively small and MOS transistors are essentially unidirectional between their gates and their remaining terminals. Strong global feedback through a series of clusters, like that found in latches or ring oscillators, simply limits the time windows that can be considered efficiently to the time delay for a signal to pass once around the feedback loop.

The Waveform Relaxation algorithm tends to grow in computation more slowly with the size of the circuit than general circuit simulators do. The rate of convergence is not strongly related to the size of the circuit because each block is only analyzed until it reaches convergence. As a circuit grows, the number of clusters increases linearly, while the size of each remains roughly constant. As a result, relaxation methods tend to scale linearly while general algorithms, with matrices growing faster than linearly, do not.

2.3 Trading Accuracy for Speed

The goal of circuit simulators, including the MOS timing simulators mentioned in the previous section, is to produce a fairly accurate prediction of the electrical behavior of a circuit. There is only a certain amount of efficiency that can be gained by restricting such analysis to a particular class of circuits. Even digital MOS simulators are not feasible for very large circuits. The remaining technique to improve speed, beyond further optimization of the present simulation algorithms, is to sacrifice accuracy.

One way to look at the speed-accuracy trade-off is illustrated in figure 2-2. In the horizontal dimension, the accuracy of the basic circuit model is varied. In the vertical dimension, a bound on the accuracy of the analysis is varied. Conventional simulation programs, including those with very rough models such as logic simulators, all lie on the horizontal axis in this picture. Point A represents a sophisticated SPICE model while point B represents a slightly simplified model used in a timing simulator. Two dimensions are used to represent levels of accuracy because unmeasured uncertainty and bounded uncertainty have different implications. The diagonal lines represent rough contours of constant accuracy. A loose bound on an accurate model might contain the same expected uncertainty magnitude as a tight bound on a rough model.

In the graph of figure 2-2, point C represents an approximate simulator. Point D represents a bounding algorithm for a rough circuit model. Examples of previous work that fall into these two regions are discussed in this section. The primary goal of this book is to derive and explore algorithms in the vicinity of point E in the graph.

2.3.1 Approximating Simulators

In an effort to do very fast simulations of large chips, programs have been developed that use vastly simplified circuit models. The extreme example of this approach is a logic simulator, or in the case of MOS logic, a switch-level

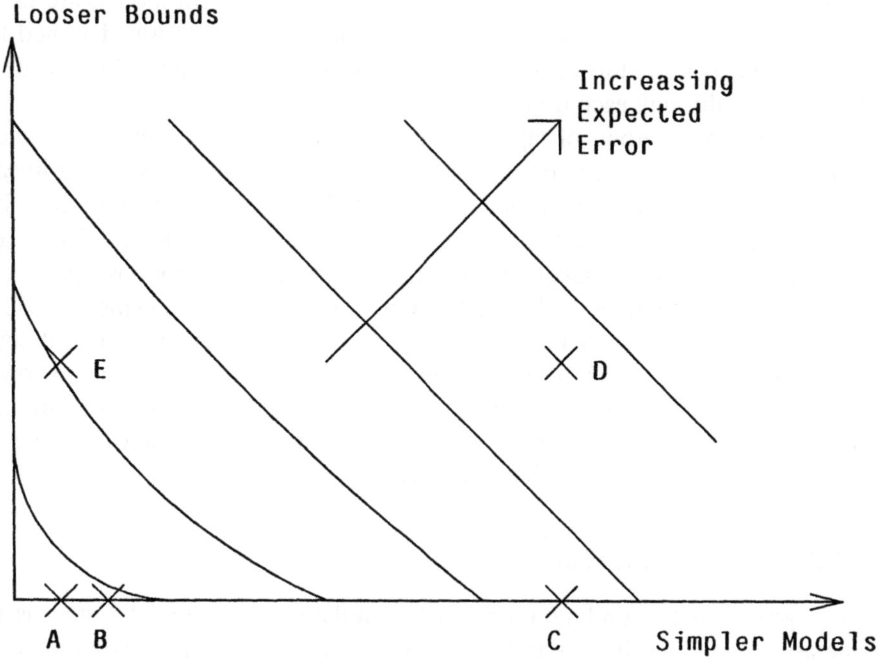

Figure 2-2: Two orthogonal approaches to the speed-accuracy tradeoff.

simulator such as MOSSIM [14] [15] [16]. Such programs completely eliminate timing information below the level of clocking. A MOS transistor is modeled simply as a voltage controlled switch. Switch-level simulators that maintain some notion of timing behavior in the logic circuits often use linear device models exclusively.

The program RSIM [17] [18] is a popular example of an approximating simulator. Intervals of linear resistance are used to model transistors, and different values are used in different circuit situations such as charging or discharging. Circuits designed using simple Mead-Conway [19] methodologies can be simulated within about 30% of their actual delay with this method. The resistance intervals used in RSIM to handle uncertainty represent the beginnings of a bounding approach, although the bounds start with extremely rough models.

Timing analyzers such as CRYSTAL [20] and TV [21] also use simple RC models for VLSI circuits. CRYSTAL searches for critical paths given user specified directions for operation of pass transistors. Different models are used for rising and falling transitions. TV is a similar program that was designed to verify circuits designed using a particular clocking methodology. The direction of signal flow through each transistor is determined by the program.

There are a number of approximate simulators being developed with accuracies that fall somewhere between exact simulators and those using linear RC models. One approach has been to develop fairly accurate macromodels for common subcircuits such as logic gates [22]. Another has been to use piecewise-linear device models [23] [24] [25]. The use of coarse discretizations for circuit variables, along with table-lookup delay functions, has also been explored [26].

Approximate simulators are very useful for first pass design and rough design verification. They can find large timing errors efficiently, but the unknown nature of their error makes them poor verifiers for high performance designs. A circuit designer currently requires a set of simulators that spans a wide range of "typical" errors in estimates of circuit behavior.

2.3.2 Bounding Approaches

The goal of the bounding approach to simulation of a general system is to generate a bound on the system response that requires less computation than an exact solution. The approach has been applied successfully in the past to simple classes of circuits, and four examples of this are provided below. Bounding algorithms derived for more general classes of circuits, though, have not tended to produce tight bounds.

One general approach has arisen from the study of the effects of roundoff and convergence behavior in general systems [27]. The notion of a sensitivity "measure" for a matrix has been derived for this purpose [28]. This technique has been used to calculate errors for small circuits, but it becomes impractical for those on the scale of VLSI. Computation scales poorly with circuit size and since it is a global technique, the worst case uncertainties grow with circuit size. Very simple general algorithms are poor because they throw away information about the detailed operation of a circuit. Another simple global bounding algorithm derived for bipolar circuits [29] exhibits delay uncertainty of an order of magnitude for circuits containing only a few transistors.

The first example of a specialized bounding algorithm concerns the behavior of linear RC tree circuits. Such networks are often used to model VLSI interconnect, and in the case of approximating simulators, can model entire circuits. They consist of a tree structure of resistors that emanates from an input voltage source and contains a grounded capacitor at each node. It can be shown that, for a step response, such a network always exhibits a voltage as a function of resistance from the input that is convex. Based on observations of this type, closed-form bounds on the step response of such a circuit have been derived [30]. The original results have recently been improved and extended to more general linear RC mesh circuits [31] [32] [33].

The second example concerns the behavior of nonlinear RC tree circuits. Often linear RC trees are too inaccurate as an interconnect model and nonlinear elements are used. By analyzing the trajectory of the solution of such a nonlinear circuit, conditions can be found where the response is monotonic with respect to the characteristics of some elements. In these cases, nonlinear elements can be replaced by appropriate linear ones to produce a linear network that exhibits bounding behavior with respect to the original circuit [34]. A circuit transformation that allows consideration of nonlinear RC trees without these powerful monotonic properties has also been derived [35].

The third example concerns the behavior of an interconnect network containing pass transistors. For an interconnect tree where all branches are pass transistors with special approximating characteristics, a transformation of variables can be performed to convert the problem back into a simpler case for which closed-form bounds can be generated. Such a conversion can be used directly to generate bounds on the behavior of some models for pass transistor networks [36].

The final example concerns the behavior of a simple MOS inverter model driving a capacitive load. By bounding the large signal output resistance of the inverter over the entire simulation period, the current flowing into the output capacitor as a function of capacitor voltage can be bounded. As a result, simple exponential bounds can be derived for the output voltage [37].

3. SIMULATION WITH BOUNDS

There is a well known law of nature, attributed to someone named Murphy, that is often cited in jest to explain unfortunate events. By this law, the most pessimistic predictions of the future always come to pass. Although this law may appear to have some validity at times, due to human psychology, fortunately our world is not exclusively governed by it[2].

When an attempt is made to generate rigorous bounds on anything, one enters a domain in which Murphy's law reigns supreme. The most pessimistic prediction is, by definition, the desired result. The worst case error is always used. The effects of errors never cancel, but instead they always accumulate. Simplifications always produce the worst possible effect. As a result, to be useful bounding must be undertaken with great care.

There is reason to believe, however, that bounding techniques are feasible in digital VLSI circuits. Rough bounds have been used successfully in digital circuits since the appearance of the first integrated logic gates. Data sheets have always listed minimum and maximum times required to produce "valid" outputs. Logic designers have routinely used these to generate meaningful bounds on the behavior of their circuits. While care must still be taken in useful bound generation for more complex circuit models, there is reason for optimism in this pessimistic domain, at least when considering digital circuits.

There is no fundamental reason why bounds can not be made arbitrarily tight in the absence of uncertainty. For a problem such as digital MOS circuit

---

[2] The completion of this book serves as a counterexample to Murphy's law [38].

simulation, where the output is not highly sensitive to input uncertainty, small simplifications can theoretically produce very tight bounds. In practice, though, efficient bounding algorithms are often difficult to generate. A simple example can illustrate why some calculations are more difficult to bound efficiently than others [39]. Consider first the calculation $Y=X-X$, with an output Y that is very insensitive to uncertainty in the input X. If the variable X is known to lie in the interval [0,1], a straightforward bound on the subtraction operator will conclude only that Y must lie in the interval [−1,1]. By ignoring any "correlations" between the two operands of the subtraction, bounding is made feasible but information is lost. Here the practice of ignoring correlations amplifies uncertainty. A calculation that does not exhibit a correlation problem is $Y=X+X$. Knowledge that the two operands must be identical does not change the conclusion that Y must lie in the interval [0,2] if X lies in [0,1].

When bounding is used at the level of each arithmetic operation in a general algorithm, the results are often disappointing due to the effect of correlations. When digital circuits are considered at the level of logic signal waveforms, the correlations that are ignored in an efficient analysis do not generally cause major problems. The effect of correlations is considered in detail throughout this chapter. For the most part, the adder example serves as a better analogy for digital MOS simulation than the subtractor. Hence, efficient bounding algorithms are possible. While the correlation problem is manageable for digital circuits, bounding algorithms should be designed with the important correlation effects in mind.

This chapter serves as an overview of the use of bounds in digital MOS simulation, including the specific approach to bounding taken in this book. Both theoretical and algorithmic details are postponed until later. In the first section useful bounds are defined for circuit behaviors, circuit models, and circuit inputs. The second section examines the bounding strategy proposed in the book. The third section considers the performance of the bounding strategy on various types of MOS subcircuits. The fourth section presents applications for a bounding simulator such as uncertainty management and worst case analysis, and demonstrates that a bounding approach is becoming essential for low-cost, reliable simulation.

3.1 Bound Definition

In an abstract sense, a bound defines a subset of a larger set in which an element must lie. Consider the set of all positive real numbers, representing possible distances from the sun. Unless the Earth's orbit has changed significantly since the writing of this book, the distance from the Earth to the sun in miles is contained in the subset consisting of all real numbers between 91 and 95 million. The lower and upper "bounds" of 91 and 95 million define a boundary between the subset and the complete set.

There are three different sets that concern circuit simulation: the set of all possible circuit excitations; the set of all possible circuit models; and the set of all possible circuit behaviors. An exact simulator returns an element in the set of behaviors when given an element from the set of excitations and an element from the set of models. A bounding simulator operates on subsets of these elements, specified by their boundaries.

The three sets just mentioned are generally complex and have an infinite number of dimensions, though they are represented in the next figure as the points inside a box. While this simple representation can be misleading, the two dimensions that paper allows are sufficient to convey many important ideas, such as the lack of a linear ordering of elements with more than one dimension. Figure 3-1 uses the simplified representation to compare exact and bounding simulators.

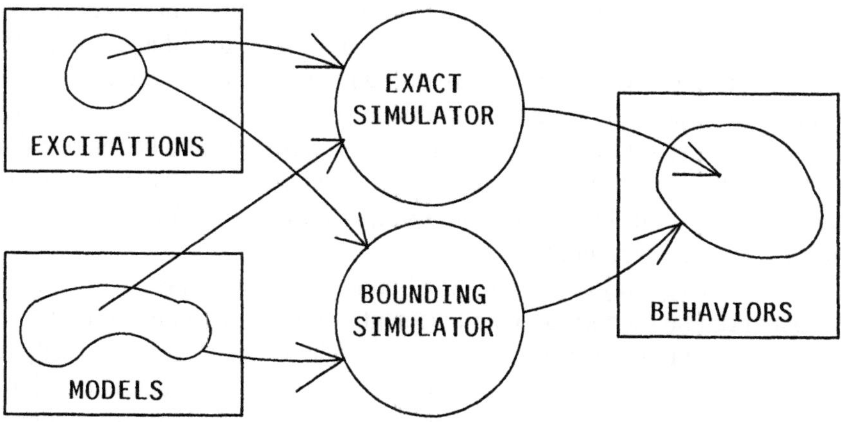

Figure 3-1: Exact and bounding simulators involve mappings between sets.

A bounding simulator is defined in terms of a truly exact simulator. If an exact simulator were used with all inputs in the excitation subset, in all possible combinations with the models in the model subset, it would be guaranteed to produce only behaviors inside the behavior subset produced by the bounding simulator. Of course, the behavior subset produced by the bounding simulator can also include behaviors that could not be produced by the exact simulator. The presence of extra behaviors is caused by two effects, called *conservative bound specification* and *conservative bound generation*, as defined below. Conservative bound specification arises from the types of bounds used, as discussed in this section, and conservative bound generation arises from the strategy used for calculating bounds, as discussed in the next section.

> **Definition 3-1**: Conservative bound specification adds elements to a subset for the purpose of specifying a bound on the subset in a more convenient form. This can apply to any of the three subsets pictured in figure 3-1. Conservative bound generation produces more than the minimum subset of behaviors that are possible for a given set of inputs to simplify computation. Both of these effects together are referred to as bounding simplification.

3.1.1 Bounding General Subsets

The abstract definition of a bound is conceptually useful, but practical aspects of specifying bounds must be considered in deriving a useful algorithm. If an exact simulator could be run an infinite number of times, it could be used to produce an "optimal" bounding simulator by forming the minimum valid behavior set. Even if this were possible, however, there would be the remaining problem of specifying the boundary of such a subset. In practice, only subsets whose boundary can be specified in a simple form are useful. This difficulty leads to *conservative bound specification.*

There are many ways to specify the boundary of a general set. An explicit expression can be used to specify all elements, or a metric can be defined on the set and the notion of distance from a single element can be used, e.g., considering only balls of varying radius. A very convenient and common technique is to define a partial ordering on the set and use two endpoint elements to specify the subset of all elements between them in the partial ordering. These interval subsets can be easily extended by including intersections and unions of intervals as well. The main advantage of the interval method is that monotonic relationships with respect to the partial ordering can be easily exploited. Often one can consider only the endpoint elements in computations. In addition, since some elements are generally easier to use in computations, these can be used exclusively as endpoints if they are spread throughout the partial ordering.

Consider, for example, two-dimensional Euclidean space, i.e., all points (x,y) where both x and y are real numbers. A simple partial ordering can be defined by considering each dimension separately. One element is said to be larger than another if it is larger in each dimension (x and y). Now consider the subset *A* defined as all points on the line y=x for 1<x<1.8, as pictured in figure 3-2. The smallest interval with respect to the partial ordering is defined by the endpoints (1,1) and (1.8,1.8). The resulting bound, represented by the small square in figure 3-2, is conservative because it contains points such as (1,1.8) that were not in the original subset *A*. The conservatism is a result of the correlation between dimensions that is ignored in the partial ordering. If only integers were used as endpoints to simplify storage and computation, the interval of [(1,1),(2,2)], also pictured as a square in figure 3-2, would be used. Simplifying the endpoints produces bounds that are even more conservative.

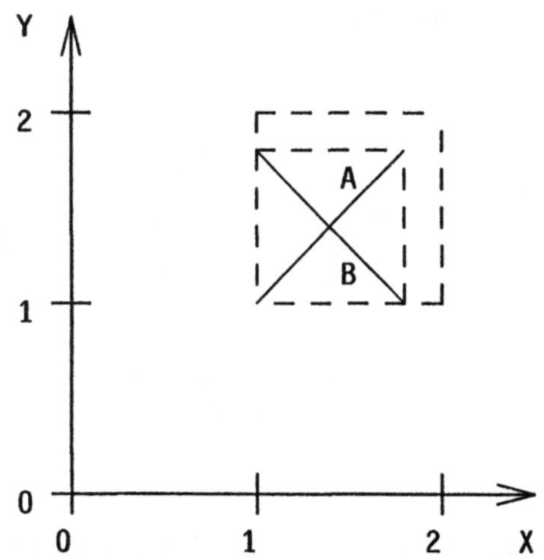

Figure 3-2: Sets can be bounded using intervals based on a partial ordering.

To determine the effect of *conservative bound specification* on a particular set, one must consider how the set is interpreted. When sets are viewed in the context of the functions that operate on them, *conservative bound specification* is of greater concern in some situations than in others. The elements that are added to a subset when a conservative bound is used may or may not give rise to more extreme and therefore more conservative outputs in a given problem. If elements in the original subset produce the most extreme outputs, using the conservative bound does not produce any more uncertainty in the output than using the original subset itself. *Conservative bound specification* is measured in terms of its effect, i.e., it is said to be large only if it leads to more conservative results.

The effects of *conservative bound specification* on the three sets considered in circuit simulation are shaped by their individual roles. The excitation and circuit model sets are used as simulation inputs, and the circuit behaviors that they give rise to are of importance. If adding an element to the set of excitations does not generate an additional element in the set of circuit behaviors, it does not affect the tightness of the resulting bounds on the solution. The behavior set is generally used as an input to a function that decides if circuit specifications are met, so the conclusions that it gives rise to through the specification function, pictured in figure 3-3, are important. If adding an element to the set of behaviors does not affect the conclusion of whether the circuit meets its specifications, the addition is harmless.

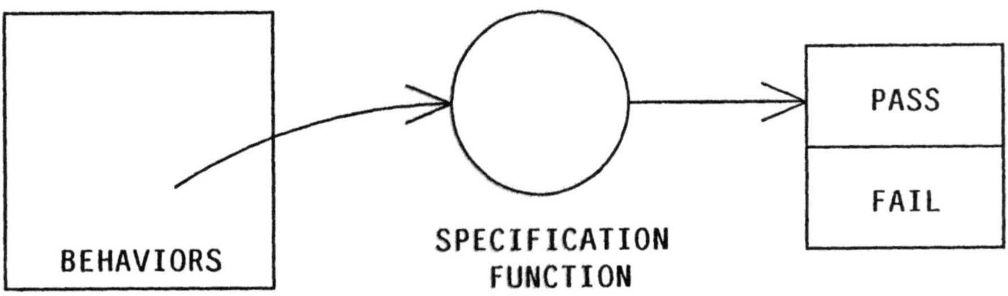

Figure 3-3: The behavior subset is interpreted through a specification function.

Consider again the example in figure 3-2, with the additional subset $B=\{(x,y)|y=2.8-x,\ 1<x<1.8\}$. The interval that is pictured, [(1,1),(1.8,1.8)], is the optimal bound for both subsets *A* and *B* given the partial ordering used. Also consider the two output functions $x+y$ and $x-y$, each operating on one of the subsets *A* or *B*. The output function $x+y$ ranges over the interval [2,3.6] when operating on subset *A*, but produces only 2.8 when operating on subset *B*.

The output function $x-y$ produces only the value 0 when operating on subset *A*, but produces values within the range $[-.8,.8]$ when operating on subset *B*. When the conservative bound for both subsets *A* and *B*, represented by the small square in figure 3-2, is operated upon by the two output functions, $x+y$ produces the range [2,3.6] while $x-y$ produces the range $[-.8,.8]$. Using the conservative bound on subset *A* only increases uncertainty when the set is viewed "through the lens of" the output function $x-y$, while the opposite is the case for subset *B*.

The next subsections describe the methods used here to bound the subsets considered in circuit simulation. Only intervals based on simple partial orderings are explored. The partial orderings used are dictated by the monotonic properties of the MOS circuit model presented in the next chapter. An interval is defined in the context of a partial ordering as all elements that are less than or equal to an upper bound, as well as greater than or equal to a lower bound. The bounds must be elements of the entire set on which the partial ordering is defined. In each case the effects of *conservative bound specification* are discussed.

3.1.2 Bounding the Excitations

The excitation of a circuit, as defined here, includes both the waveforms of a set of time-varying, grounded voltage sources that drive the circuit, as well as the circuit's initial state. Voltage sources can be used, along with passive devices in the circuit model, to represent most real inputs accurately. The power supply signals are treated as constant inputs.

Both the value and the time derivative of each input voltage source are generally relevant to the operation of MOS integrated circuits. Since distinct monotonic relationships exist for each, they must be specified independently to produce a useful partial ordering. The separation of derivative information creates the interesting situation that some excitations that satisfy the definition here are physically impossible. Of course, only the elements that have consistent voltages and voltage derivatives can reflect real excitations.

The initial state of a circuit is also specified as part of its excitation to represent the effect of the input voltage sources before the initial time in a simulation. The voltage source waveforms need only be specified during the window of time considered in a simulation. A standard assumption for an integrated MOS circuit model is that it contains no inductance so its state is determined solely by its capacitor voltages. As a result, the initial state is specified by a number of initial voltages at internal nodes of the circuit.

Definition 3-2: An <u>excitation</u> is a pair $\{u,s\}$ where $s\in\mathbb{R}^{n-m}$ specifies an initial voltage on a subset of internal nodes in some network $\mathcal{N}$ representing the initial state, and $u:[0,T]\rightarrow\mathbb{R}^{2m}$ is a continuous function that maps values of time into real values, and "time derivatives"[3], of m input voltage sources.

The partial ordering of excitations is defined by partitioning the coordinates of the excitation set in five different ways: among the voltage sources, between values and derivatives, among points in time, between input waveforms and initial states, and among the initial internal node voltages. This partitioning is analogous to the partitioning of the coordinates x and y in figure 3-2. Two excitations are ordered if they are ordered in each of the one-dimensional projections generated by the partition. Partitioning the coordinates of the excitation set leads to *conservative bound specification.*

Definition 3-3: Given two excitations $\{u,s\}=\{v_0(t), \dot{v}_0(t), v_1(t), \ldots \dot{v}_{m-1}(t), v_m(0), \ldots v_{n-1}(0)\}$ and $\{\hat{u},\hat{s}\}=\{\hat{v}_0(t), \dot{\hat{v}}_0(t), \hat{v}_1(t), \ldots \dot{\hat{v}}_{m-1}, \hat{v}_m(0), \ldots \hat{v}_{n-1}(0)\}$, we say that "$\{u,s\}\geq\{\hat{u},\hat{s}\}$" iff a) $v_i(t)\geq\hat{v}_i(t)$ and $\dot{v}_i(t)\geq\dot{\hat{v}}_i(t)$ $\forall$i such that $0\leq i\leq m-1$, and $\forall t\in[0,T]$, and b) $v_i(0)\geq\hat{v}_i(0)$ $\forall$i such that $m\leq i\leq n-1$.

It is possible at this point to do a rough analysis of the effect of *conservative bound specification* for excitations in digital MOS simulation. The extremes of behavior in a digital MOS circuit correspond to, for the most part, fast and slow transitions. Along with some intuition about timing in MOS circuits, this observation can be used to determine which ignored correlations in the excitation subset might have a significant impact on the resulting behavior subset. Each of the five types of partitioning for the excitation set is a source of *conservative bound specification*, and each is analyzed separately here.

---

[3]The quotes emphasize the fact that, while the time derivative plays its normal role in the circuit, it need not actually be the derivative of the corresponding signal in the excitation.

The first source of *conservative bound specification* is the partition between the input waveforms and the initial state. When doing simulation, circuit designers often desire to have the initial state correspond to a d.c. solution of the circuit when the input voltage sources are held fixed at their initial values. As long as the initial inputs are valid logic values and the initial state consists of the valid logic values observed when the circuit is driven by the initial inputs, the effect of ignoring this correlation is small. The maximum distance of any initial node voltage from the d.c. value corresponding to a valid set of inputs is only the width of the range of valid logic voltages. In addition, even if there are significant differences, they fade quickly in percentage terms over time because they are one-time occurrences.

The second source of *conservative bound specification* for excitations is the partition among the initial node voltages. If the actual initial state is a d.c. solution of the circuit, only certain combinations within an interval based on the partial ordering are possible. As with correlations between the initial state and the inputs, the effect of correlations among components of the initial state tends to be small and fade quickly over time. For a single nonreconvergent critical path through a combinational logic circuit, this correlation can have little effect even when uncertainty in the initial state is large. Consider a simple critical path consisting of a chain of two inverters driven by a falling waveform. If the initial values of the internal nodes are completely unknown, i.e., they have bounds of [0, VDD], the slow case will start with the output of the first inverter low, and the output of the second inverter high. The fast case will be exactly opposite. The inconsistent cases of both voltages being high or both being low do not create overly conservative extremes of behavior. Using more complex logic gates[4] or additional logic stages does not affect this lack of sensitivity to correlations among initial voltages.

The third source of *conservative bound specification* is the partition among the individual input voltage sources. Even when there are large uncertainties in the inputs, this simplification is not a great concern. Digital circuits rarely depend on strong correlations between the arrivals of inputs to function correctly. The output of a digital circuit generally switches earliest when all its inputs switch earliest. The slow-fast "corners" where some inputs are fast (early) and others slow (late), corresponding by analogy to the point (1,1.8) in figure 3-2, might never occur but do not generally produce extra behaviors outside the range of

---

[4] These logic gates must have d.c. output voltages that are roughly monotonic functions of their input voltage vectors, e.g., NAND gates but not EXCLUSIVE-OR (XOR) gates.

those produced if the inputs are all fast or all slow. In other words, this expansion of the input subset does not produce a significant expansion in the corresponding behavior subset.

The fourth source of *conservative bound specification* is the partition between a signal and its time derivative[5]. The next chapter shows that the derivative is only of concern for inputs that are coupled to other circuit nodes through capacitors. Even in these cases, internodal coupling is generally small in digital circuits and behavior is not very sensitive to small shifts in input derivative waveforms. This ignored correlation becomes a concern only when very tight bounds on gates with Miller capacitance are desired, or when special subcircuits are considered that use the correlation between signals and their derivatives for correct operation.

The final and most important source of *conservative bound specification* is the partition among values of a signal at different times. This partition has different implications for voltage waveforms and voltage derivative waveforms. Partitioning voltage waveforms among points in time has little effect, as with partitioning among input voltage sources. The unlikely voltage waveforms that follow a high extreme for some times and a low extreme for others do not significantly expand the output behavior subset. For inputs coupled to other nodes through capacitors, though, the bounds used for time derivative waveforms can inject significant uncertainty. Consider the set of input waveforms for one voltage source pictured in figure 3-4. The waveform can start rising from 0 volts at any time between 0 and 35 ns, and then rise to 5 volts with a slope of 1 v/ns. Four possible waveforms are pictured in the figure. The best (smallest) upper bound on the time derivative, based on the partial ordering, is 1 v/ns over the entire time between 0 and 40 ns. While the time integral of the actual derivative waveform must be 5 volts, the upper bound has a time integral of 40 volts. Since the time integral in part determines the amount of charge pushed through any internodal coupling capacitors into other nodes, the partition among points in time can have a significant effect on sensitive circuits, e.g., those containing dynamic storage nodes, even when coupling capacitors are quite small. In cases where this effect is significant, additional information must be used to bound total charge sharing between nodes. One potential solution to this problem is the use of charge as a state variable instead of voltage, at least for some capacitors. For this reason, chapter four treats both possible choices of state variable.

---

[5] Recall that a signal waveform has a unique derivative, and the two are therefore strongly correlated.

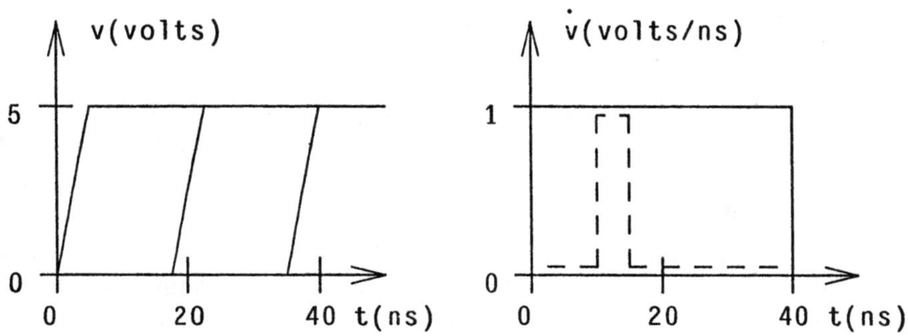

Figure 3-4: Ignoring correlations over time can produce unattainable bounds.

3.1.3 Bounding the Models

The circuit models considered here consist of networks of lumped element models. Models for digital MOS integrated circuits generally contain elements that correspond to transistors, resistors, and capacitors, along with input ports. A detailed definition of the network elements and element composition rules is postponed until the next chapter. At this point only a rough definition, that of some lumped element network containing the three device types, is used for the circuit model.

The partial ordering of models is defined by partitioning the model set among the individual elements. A comparison of two models is made in an element-by-element fashion, using the partial orderings defined for each element type. The orderings of each element type are considered in chapter four along with their definitions. Transistors and resistors are ordered according to their current magnitude at each combination of terminal voltages, while capacitors are ordered according to their incremental capacitance at each voltage. To be ordered, two models must be identical except for the constitutive relations of their elements.

Definition 3-4: Given two networks $\mathcal{N}$ and $\hat{\mathcal{N}}$ with the same topology, consisting of elements $\{E_1, E_2, .. E_p\}$ and $\{\hat{E}_1, \hat{E}_2, .. \hat{E}_p\}$ respectively, we say that " $\mathcal{N} \geq \hat{\mathcal{N}}$ " iff $E_i \geq \hat{E}_i$ $\forall$i such that $1 \leq i \leq p$.

The *conservative bound specification* for models arising from the partial orderings used for circuit elements is not important because it does not significantly expand the behavior space. The partial orderings used for elements generally produce bounds that, in addition to being possible elements themselves, also give rise to fast and slow extremes of circuit behavior. Unlikely transistors with larger than average currents at some operating points and smaller than average currents at others do not produce extreme behavior in restoring logic circuits. The partitioning among individual circuit elements, however, can have a significant effect.

In integrated circuits there is often a large correlation between elements that appear in the same circuit. Even though digital circuits do not generally rely on these correlations to operate correctly, disregarding them can lead to overly pessimistic performance bounds in the presence of large uncertainties in device parameters. One way to avoid this problem is to consider the model in terms of lower-level parameters, such as mobilities, that reflect the source of strong correlations. Such an approach is not explored in this book as the primary goal here is not incorporating large uncertainties in device parameters, but rather incorporating small device simplifications. Applications of bounding algorithms are discussed further in the fourth section of this chapter.

3.1.4 Bounding the Behaviors

The behavior of a circuit can be characterized in many different ways. The best method depends in large part on the function performed by the circuit and how the circuit's behavior is calculated. The behavior of an oscillator might best be described by a frequency of oscillation, an amplitude, and a duty cycle. The behaviors of the digital circuits considered here are best described by a set of node voltage waveforms. The node voltages are appropriate because they are used to represent the logic signals being processed by the circuit and because they generally specify the state of the circuit at each point in time. Node voltages are also appropriate to use when exploiting the monotonic properties of MOS subcircuits. As with excitations, the time derivatives of the node voltages must be treated independently for use in monotonic relationships.

Definition 3-5: A <u>circuit behavior</u> is a continuous function $w:[0,T] \rightarrow \mathbb{R}^{2(n-m)}$ that maps values of time into real values and "derivatives"[6] of $n-m$ internal node voltages (state variables) of some network $\mathcal{N}$.

The partial ordering used for behaviors is identical to that used for excitations if input voltage sources are replaced by node voltages and if the initial state is ignored. As a result, the sources of *conservative bound specification* in behaviors are similar to those found in excitations.

Definition 3-6: Given two circuit behaviors $w=\{v_m(t), \dot{v}_m(t), \ldots \dot{v}_{n-1}(t)\}$ and $\hat{w}=\{\hat{v}_m(t), \hat{\dot{v}}_m(t), \ldots \hat{\dot{v}}_{n-1}\}$, we say that " $w \geq \hat{w}$ " iff $v_i(t) \geq \hat{v}_i(t)$ and $\dot{v}_i(t) \geq \hat{\dot{v}}_i(t)$ $\forall i$ such that $m \leq i \leq n-1$, and $\forall t \in [0,T]$.

There are two different types of components in a circuit behavior. The first corresponds to part of the circuit that serves as an input to another portion. Due to the primarily one-way coupling between digital subcircuits, these components have the same context as circuit inputs. The second type of behavior component corresponds to outputs of the entire circuit. These components are considered in the context of various performance specifications for the circuit. The specifications are usually formulated with the assumption that the circuit outputs will be used as inputs for another digital circuit, though, so the context of these components is also similar to that of circuit inputs. Since behaviors and excitations have similar correlations and contexts, the analysis for excitations, minus the consideration of initial state, is valid for behaviors as well.

Even though *conservative bound specification* is not usually of major concern for behaviors, it can be disconcerting at times. An extreme example is a chain of two inverters in which one possible behavior has the output of the second switching before its input does. In this "corner" of the behavior subset,

---

[6] Again the quotes indicate that corresponding time derivatives and signals need not match in a circuit behavior, even though the time derivative plays the role of a time derivative in the circuit.

analogous to the point (1, 1.8) in figure 3-2, the output of the first inverter is at its slowest and that of the second is at its fastest. If both of these signals are used as inputs to a logic gate, though, the extreme cases occur when both signals are fast and when both are slow, and the implausible but allowed behaviors have no effect.

3.1.5 Forms of Bounds

Just as the bounds in figure 3-2 can be simplified by restricting them to be integers, the bounds used for excitations, models, and behaviors can be restricted to simple forms. The form has implications in storage requirements, computation, and accuracy. By introducing a little simplification in a bound specification, large savings in storage and computation are possible without large increases in uncertainty.

When circuit models are bounded, there are two levels at which simplifications can occur. First, element constraints can be expressed with linear or piecewise-linear functions. The number of breakpoints used in piecewise-linear functions often provides a simple mechanism for trading accuracy for speed. Second, elements can be simplified as a group so that their forms can be easily combined when larger circuits are considered. For example, two parallel resistors with similar bound forms, either with breakpoints at the same voltages or with scaled versions of the same i-v curve, can be efficiently merged into one resistor for purposes of computation.

Similar techniques, such as the use of piecewise-linear waveforms, are possible for both excitation and behavior bounds. By simplifying voltage bounds as a group, one might produce a situation where two parallel transistors have identical signal bounds at their gate terminals, allowing the two to be efficiently combined in computation. In addition, calculations can sometimes be greatly simplified by using waveforms that are monotonic in time, and real waveforms often do not have this property. Fortunately, any bound on a restored logic waveform can be simplified during one clock cycle into two distinct monotonic bounds; one that is good for small times (before the first transition) and one that is good for large times (after the last transition). Figure 3-5 illustrates how a nonmonotonic upper bound on a single waveform can be transformed into two valid monotonic ones. For optimal monotonic bounds, one bound[7] has as its value at each time t the supremum (least upper bound) of

---

[7] If the original bound is a continuous function, the optimal monotonic bounds as defined here are as well.

the original bound over the entire interval [0,t]. The other waveform bound uses the supremum (least upper bound) over the interval [t,T]. The monotonic bounds pictured in figure 3-5 are slightly more conservative than the optimal ones. At most times one of the bounds is poor, but only during transitions between logic levels can both fail to be tight, e.g., during the intermediate dip in the waveform in figure 3-5. A symmetric technique can be used for lower bounds. Considering the upper and lower bounds together, they define two intervals whose intersection contains the actual waveform. The extra computation involved in considering a bound expressed as the intersection of two independently valid bounds, approximately a factor of two, can be efficient if that amount is saved by the use of monotonic waveforms. In addition, in special applications such as a worst case timing analyzer, often only the bound that is tight for large values of time is important.

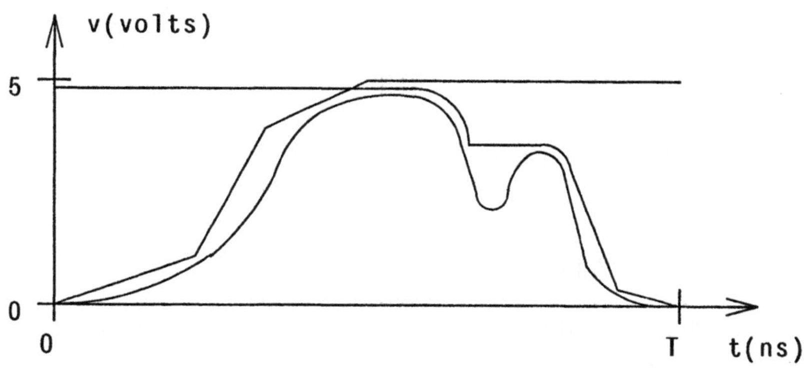

Figure 3-5: Nonmonotonic bounds can be simplified into monotonic ones.

The forms that are used for bounds on behaviors can range from waveforms characteristic of logic simulators all the way to those characteristic of exact electrical simulators. In one sense, bounds can serve as a natural bridge between these different levels of circuit analysis. Figure 3-6 illustrates an extremely simple form for a bound on a voltage waveform that corresponds to a 0-X-1 characterization. Such a characterization is often used in high-level switch-type simulators such as MOSSIM [14], where each node is either guaranteed to be a logic "1" or "0", or is represented by an unknown "X" at a given time. A voltage is said to be a logic "0" if it is guaranteed to lie below some threshold voltage V_{IL} and a logic "1" if it is guaranteed to lie above some threshold voltage V_{IH}. These 0-X-1 bounds can be improved to a certain extent by modifying only the X portion to obtain accuracy, while still exploiting the restoring nature of logic and confining calculations to transition periods.

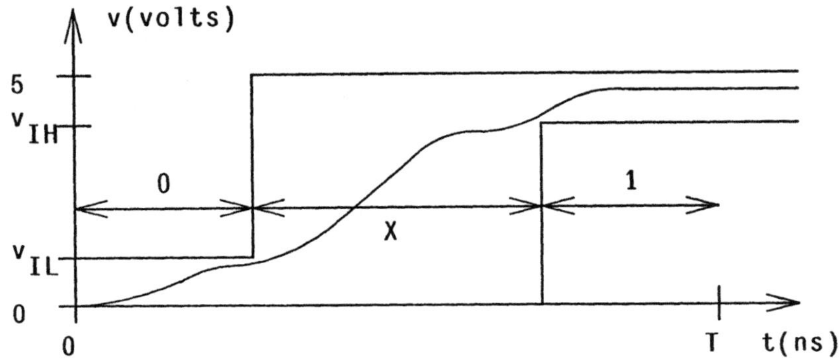

Figure 3-6: Logic simulation waveforms can be represented with bounds.

3.2 Basic Strategy

The basic strategy used in this book to derive bounding algorithms has three distinct levels. At the lowest level, highly efficient methods are used to analyze very simple circuits and situations. The second level involves rigorously transforming more realistic models, at least for small subcircuits, into the simpler problems tackled at the lowest level. This transformation is achieved by exploiting monotonic properties of the models. The top level uses a relaxation technique to partition analysis among small subcircuits. The theoretical results presented in the book concern primarily the middle and top levels and are considered in chapters four and five, respectively. Some algorithms that are useful at the lowest level are presented in chapter six. Here, the overall strategy is analyzed in general terms.

3.2.1 Bounding the Behavior of Simple Circuits

This subsection discusses the lowest level of the basic bounding strategy. The word "simple" in the title is not meant to imply that algorithms derived at the lowest level are trivial or unimportant. It merely signifies that the circuits and situations considered are highly simplified models of VLSI subcircuits, e.g., the linear RC line pictured in figure 3-7. A good analogy for this level is the solution of linear equations in exact simulators. A large effort is put into efficient solution because it takes place in the innermost loop of a simulation algorithm, and is therefore crucial to the efficiency of the entire algorithm.

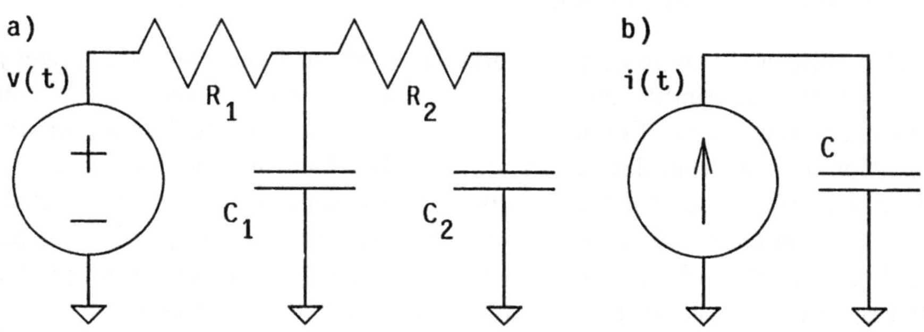

Figure 3-7: Simplified circuits are analyzed inside a bounding algorithm.

The circuits considered at the lowest level are generally linear, piecewise-linear, or contain very special nonlinearities. While they can often be solved exactly with a reasonable amount of computation, bounding techniques can sometimes be used to reduce computation further. Most previous work on evaluating and bounding the solutions of simplified circuit models presented in chapter two can potentially be used here. The use of bounding techniques at the lowest level is one source of *conservative bound generation.* Strictly speaking, to obtain truly rigorous bounds on the behavior of the complex circuit model, the simplified (bounding) circuit models must also be analyzed with rigorous bounds. Even numerical roundoff would have to be taken into account. This can be done, but beyond a certain level of confidence it becomes unnecessary. "Exact" numerical solutions for simple circuits such as the one in figure 3-7b would normally be adequate in practice.

The main general question at this level, if bounding techniques are used, involves matching the accuracy with that at higher levels. In some sense, each level should have a relationship between accuracy and computation that can be differentiated to find the marginal accuracy gained from additional computation at any operating point. For maximum efficiency, a bounding algorithm should operate at a point where the marginal accuracy is equal at all levels. If a simple circuit is used to generate a bound on the behavior of a complex circuit model with the introduction of about ten percent error, it probably does not pay to bound the behavior of the simple circuit to within 0.1 percent. Beyond a point, computation would provide a greater return if used at a higher level.

3.2.2 Bounding the Behavior of Monotonic Subcircuits

This subsection discusses the middle level of the basic bounding strategy. Given that very simple circuit models can be either solved or bounded efficiently, one would like to use these results to generate bounds on the behavior of more complex circuit models. The ideal situation is one where, given a complex circuit model, a simple transformation can be undertaken to produce a simple circuit whose behavior bounds that of the original. This can be done for small MOS subcircuits by exploiting their monotonic properties. As long as the simplified circuit maintains the essential properties of the original one, e.g., the restoring nature of digital logic, efficient bounds can be produced. In this level of the bounding strategy, only small subcircuits (corresponding roughly to logic gates) of a complete and sophisticated VLSI circuit model are considered.

A subset of the circuit behavior is associated with each subcircuit, corresponding to the voltages of each of its internal nodes and their time derivatives. A subcircuit must have at least one internal node. The behavior of the rest of the circuit is treated as a known input to the subcircuit[8]. It is assumed that given a model for the subcircuit, its initial state, and its inputs, there is a unique solution to its behavior.

As a simple example, consider the circuit consisting of the ring oscillator with three inverters pictured in figure 3-8. The circuit behavior that corresponds to this circuit consists of three node voltage waveforms $v_1(t)$, $v_2(t)$, and $v_3(t)$, along with their time derivative waveforms. Also pictured in the figure is a subcircuit

---

[8] In the context of bounding algorithms, a known input is one that is restricted to a known interval.

with only node one as an internal node. The behavior of the subcircuit consists of a single voltage waveform $v_1(t)$ and its time derivative. When considering only the subcircuit, the remaining variables in the entire circuit are treated as known inputs represented by voltage sources. The transistor models are assumed to be completely resistive in nature, as well as unidirectional between the gate and remaining terminals. All transistor capacitance is modeled with external elements. As a result, transistors T3 and T4 are not included in the subcircuit since they have no direct effect on the first node.

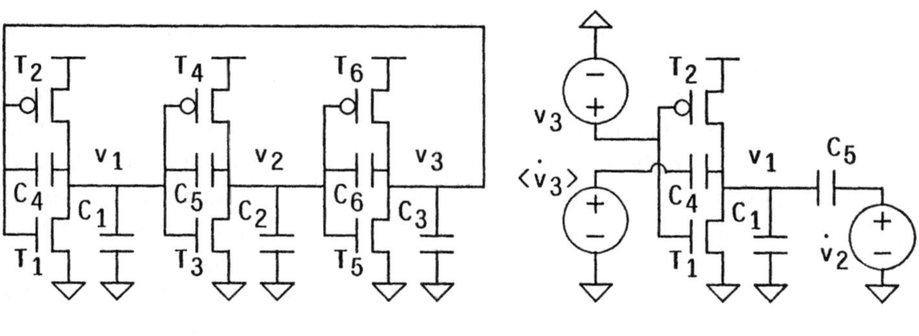

Figure 3-8: Large MOS circuits can be subdivided into monotonic subcircuits.

The next chapter shows that reasonable MOS circuit models can be divided into subcircuits with behaviors that are simple monotonic functions of both their circuit model and their inputs, where monotonicity is defined in terms of the previously mentioned partial orderings. For example[9], the voltage waveform $v_1(t)$ of the subcircuit in figure 3-8 is a monotonically decreasing function of the input waveform $v_3(t)$. Second, it is a monotonically increasing function of the input waveform $\dot{v}_3(t)$. Third, it is a monotonically increasing function of the drive of transistor T2. Similar relationships are valid for the remaining inputs and circuit elements as well.

Due to these monotonic properties, the behavior of the subcircuit can be bounded by the response of a much simpler circuit that corresponds to the endpoints of intervals bounding each input and circuit element. Uncertainty in

---

[9] The simple relationships listed here require all voltages to fall within the range of [0,VDD].

the inputs and element characteristics can easily be incorporated in this simplification process. Since the waveform $v_1(t)$ is a monotonically decreasing function of the waveform $v_3(t)$, the behavior of $v_1(t)$ when the input $v_3(t)$ is fixed at its upper bound is guaranteed to be a lower bound on the original behavior. The upper bound on $v_3(t)$ could be a piecewise-linear waveform, or it could have an even simpler form, and thus replace the original subcircuit by one that requires less computation to analyze. Since the waveform $v_1(t)$ is monotonic in the drive of transistor T2, the behavior of $v_1(t)$ is also bounded from beneath by that of a circuit where the transistor model is replaced by one with a smaller current drive. As with the input waveform, this lower bound can correspond to a linear or piecewise-linear device, and thus greatly simplify the circuit that is analyzed.

For the simple example of the inverter just presented, there is no *conservative bound generation* associated with the exploitation of direct monotonic properties. The simplification that arises from the use of simple endpoints such as piecewise-linear waveforms, was included as part of *conservative bound specification* in section 3.1. For more complex circuits, though, the behavior is not always directly monotonic with respect to each input and element characteristic. Monotonic relationships still exist that allow bounds to be computed, but some correlations must be ignored in the computation. In this case additional *conservative bound generation* is produced. A clear example of this effect from outside the realm of circuits is the calculation $Y=X^2-2X$, $X>0$. Y is not directly monotonic in X, but it is related by a composition of monotonic functions. Since the subtraction function is monotonic in each operand, Y is monotonic in each of the two terms X^2 and 2X if they are treated independently. Each term is then monotonic in X since X must be nonnegative. Suppose that X lies in the interval [0,2]. This guarantees that Y lies in the interval [−1,0], but since Y is not monotonic in X, the true range of Y cannot be computed using only endpoints. Using only the monotonic properties one concludes that $X^2 \in [0,4]$, $2X \in [0,4]$, and therefore $Y \in [-4,4]$, producing considerable *conservative bound generation.*

The specific correlations that are efficient to ignore when analyzing complex circuit models are discussed in chapter four. The most significant simplification takes place when large uncertainty exists in the current passing through a transistor between two internal subcircuit nodes, and the correlation between the effects of the current on each node is ignored. Correlations within logic circuits constitute a challenge for general bounding algorithms, but the search for more powerful solutions takes place in the small, well defined domain of small subcircuits. The basic subcircuit for a digital MOS circuit is simply a cluster of transistors that alternately charges and discharges capacitors.

3.2.3 Relaxation of Bounds

This subsection discusses the top level of the basic bounding strategy. When taken beyond the level of small subcircuits that correspond roughly to logic stages, direct exploitation of monotonic properties becomes quite complex. The partitioning of analysis that is essential for efficient treatment of large, digital MOS circuits is especially important for bounding algorithms. For simple circuit models that contain only one-way coupling between subcircuits, partitioned analysis is straightforward. Bounds on waveforms can be propagated from inputs to outputs of subcircuits using the bottom two levels of the bounding strategy along with event driven scheduling of analysis. For more complex circuit models, such as those containing Miller capacitance, more sophisticated techniques are needed to combine the separate analysis of each subcircuit. This combination is the top level of the bounding strategy.

This subsection treats circuits constructed from a number of interconnected blocks, each of which can be efficiently analyzed independently. A block could range from a very simple transistor cluster to a group of fairly complex, tightly coupled transistor clusters. A relaxation approach can be used to couple separate analysis of these individual blocks into a rigorous analysis for the entire circuit. Fortunately, not only can relaxation be rigorously combined with bounding, but it is also computationally efficient in digital MOS circuits. A very simple example, outside the domain of circuits, is provided here to illustrate the idea of bound relaxation and the *conservative bound generation* that it produces.

Consider the system illustrated in figure 3-9 that contains two coupled blocks, called Saturn and Titan, each with a single scalar input and output (or behavior). Titan is an identity element, so its behavior is strongly influenced by its input from Saturn. The behavior of Saturn, though, is not strongly influenced by its input from Titan if the magnitude of α is much less than one. With a small value of α, this system is an analogy for two clusters in a digital MOS circuit with primarily one-way coupling. The constraints on the two variables x and y that are imposed by the two blocks are graphed in the figure for a few values of α. Note that the system always has a unique solution for $\alpha \neq 1$. The gain around the feedback loop formed by the two blocks is α, so $\alpha > 0$ corresponds to positive feedback, while $\alpha < 0$ corresponds to negative feedback.

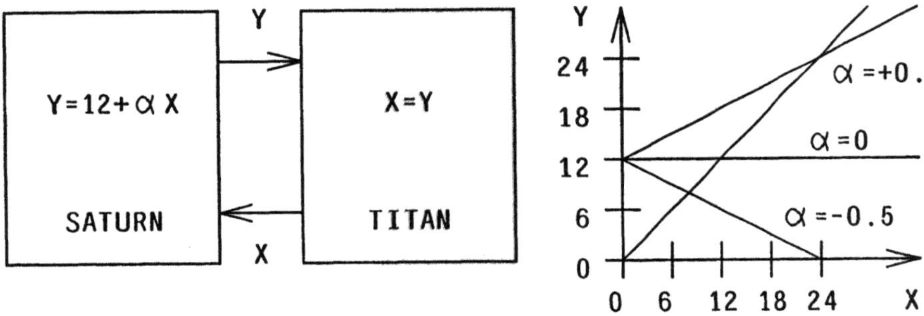

Figure 3-9: Simple example of a coupled system.

If a relaxation approach is used to find the exact solution of the system, a sequence of improving estimates $\{(x_n, y_n)\}$ is generated that converges to the solution (x_*, y_*). For this example, subscripts are used to denote iteration count. One possible relaxation algorithm[10] uses $y_{n+1}=12+\alpha x_n$ and $x_{n+1}=y_{n+1}$ to update the two variables, giving rise to the iteration equation $x_{n+1}=12+\alpha x_n$ for the variable x. The resulting sequence for x is $x_n=(12/(1-\alpha))+(x_0-12)\alpha^n$, which converges to the solution $x_*=(12/(1-\alpha))$ for any x_0 as long as $|\alpha|<1$. Convergence is only fast when $|\alpha|<<1$, i.e., the coupling between the blocks is primarily one-way. Figure 3-10 illustrates graphically how a sequence $\{x_n\}$ is produced by relaxation for $\alpha=-0.5$ and $\alpha=+0.5$.

Now consider the situation where each block is analyzed with bounds. Assume that Saturn's behavior is known to lie in the interval $[11+\alpha x, 13+\alpha x]$, and Titan's is known to fall inside $[y-1, y+1]$. These bounds restrict the possible solutions (x_*, y_*) to a diamond shaped region in the x-y plane, as pictured in figure 3-11. Since the variables x and y are considered separately in the specification of bounds, any bound on the behavior of the entire system is represented as a rectangle in the x-y plane and is therefore conservative in its

---

[10]Relaxation is actually a class of algorithms. If the latest guesses for each variable are always used in a sequential algorithm, as in the example here, it is called Gauss-Seidel relaxation.

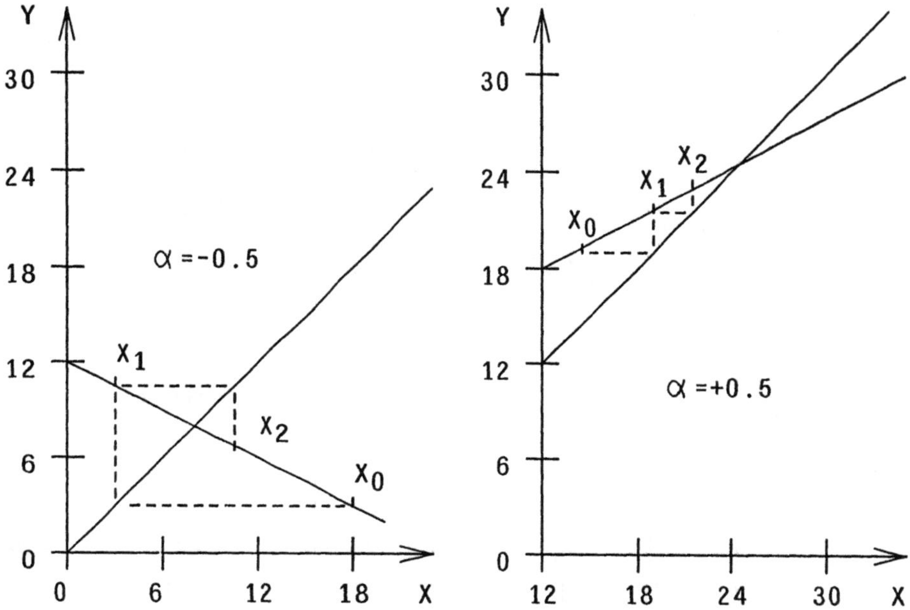

Figure 3-10: Exact relaxation towards the solution for $\alpha = -0.5$ and $\alpha = +0.5$.

specification. Depending on the sign of α, though, additional *conservative bound generation* can be introduced when separate bounds on the behavior of Saturn and Titan are combined.

The solution for the coupled bounding problem can also be found with a relaxation approach that considers intervals for the variables x and y. For Saturn, $[y_{n+1}^L, y_{n+1}^U] = [11+\alpha x_n^L, 13+\alpha x_n^U]$ if $\alpha \geq 0$ and $[11+\alpha x_n^U, 13+\alpha x_n^L]$ otherwise. For Titan, $[x_{n+1}^L, x_{n+1}^U] = [y_{n+1}^L - 1, y_{n+1}^U + 1]$. Just as in the exact case, the relaxation of intervals converges to a set of intervals that is self-consistent, i.e., the bounds on each variable give rise to the bounds on the other in separate analysis of each block.

For $\alpha > 0$, the feedback is positive and the lower bound x_n^L is used to calculate the lower bound y_{n+1}^L, which in turn is used to produce the next lower bound for x, namely x_{n+1}^L. The iteration equation for this process is

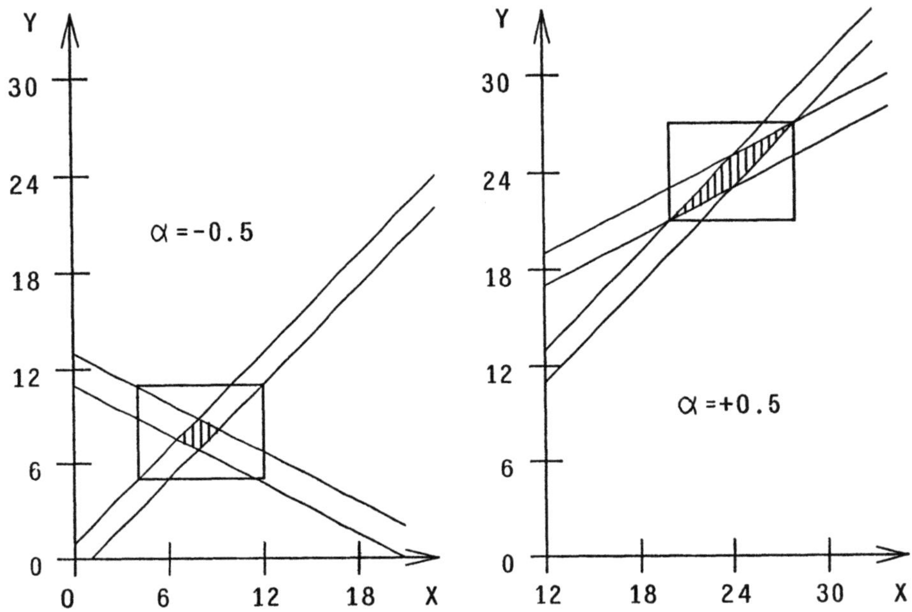

Figure 3-11: Relaxation of bounds for $\alpha = -0.5$ and $\alpha = +0.5$.

$x_{n+1}^{L} = (11 + \alpha x_n^{L}) - 1$, which converges to $10/(1-\alpha)$ with the same rate as in the exact case. Likewise, the upper bound $\{x_n^U\}$ converges to $14/(1-\alpha)$. These bounds, along with the corresponding bounds on *y*, produce the smallest possible rectangle about the region of possible solutions. Due to the positive feedback, the endpoints are possible solutions and there is no additional *conservative bound generation* introduced[11].

For $\alpha<0$, the feedback is negative and the upper bound x_n^U is used to calculate the lower bound y_{n+1}^{L}, which in turn is used to produce x_{n+1}^{L}. By

[11] In general, as long as a bound is self-consistent through all feedback paths among blocks, it is an "optimal" bound given the information provided by the bounds used for the behavior of each block.

combining two complete iterations, the iteration equation for the lower bound of x is $x_{n+2}^{L} = y_{n+2}^{L}-1 = (11+\alpha x_{n+1}^{U})-1 = 10+\alpha(y_{n+1}^{U}+1) = 10+\alpha((13+\alpha x_{n}^{L})+1) = 10+14\alpha+\alpha^2 x_{n}^{L}$, which converges to $(10+14\alpha)/(1-\alpha^2)$. Calculation of the other endpoint gives rise to the rectangle pictured in figure 3-11 for $\alpha=-0.5$. Clearly, the negative feedback has introduced a large amount of *conservative bound generation* in the coupled solution[12].

For the extreme case where $\alpha=0$, there is no feedback and no matter what the initial bounds on x are, the first iteration produces the solution. The bound on y is simply [11, 13] from the constraint on Saturn's behavior, and this implies, along with the constraint on Titan's behavior, that the bound on x is [10, 14]. Even if α is not exactly zero, this example indicates that for small α convergence is fast and *conservative bound generation* is small, if it exists at all (i.e., $\alpha<0$).

The example of figure 3-9 illustrates the main principles involved in both exact relaxation and bound relaxation. For each type of feedback, when using simple bounds on the relaxing blocks instead of exact analysis, the rate of convergence is the same and only depends on the nature of coupling between the blocks. If the coupling is primarily one-way, or small in both directions, a relaxation sequence converges very quickly. In the exact case, the sequence converges to the exact solution, while in the bounding case, it converges to a unique set of self-consistent bounds. The self-consistent bounds combine the bounds on individual blocks to produce a bound on the entire system.

The scalar example provides a good intuition for the relaxation of bounds on circuit behavior. The convergence proof for Waveform Relaxation is used in chapter five to show that behavior intervals can be relaxed, under reasonable conditions, to produce a telescoping sequence of rigorous bounds. In practice, with reasonable bounding algorithms, the rate of convergence is similar whether relaxing bounds or exact waveforms. While the self-consistent bounds that the sequence converges to are not in general guaranteed to be unique[13], in practice they have the same properties with regard to *conservative bound generation* that was illustrated in the example.

---

[12] Note that, oddly enough, it is positive feedback that is desirable in the bounding context.

[13] By slight changes in the bounds on each block that are pictured in figure 3-11 for $\alpha=-0.5$, one can form a number of distinct "rectangles" that are self-consistent. Each one is still a valid bound on the entire system, though.

There are three sources of *conservative bound generation* that can be introduced at this level of the bounding strategy. The first source only arises when negative feedback is present. As seen in the example of figure 3-9, negative feedback produces a situation where a lower bound on a variable is calculated by assuming that it is, when viewed around the feedback loop, at its upper bound. Partitioning a feedback loop causes correlations between a signal and its influence around the loop to be ignored. The second source of *conservative bound generation* at the top level arises when relaxation sequences are not taken to convergence. Intermediate bounds are rigorous, but have slack that can be removed through further relaxations. The third source of *conservative bound generation* arises when correlations among blocks in the partition are ignored. This source was not exhibited in the example because there was no correlation between the transfer functions of the two blocks. As in the circuit of figure 3-8, internodal coupling capacitors in digital MOS circuits usually appear in more than one subcircuit, producing small correlations between blocks that have an effect when uncertainty exists in the capacitor characteristics[14].

Digital MOS circuits are constructed from a set of transistor clusters. The feedback between clusters is generally negative, but is small enough that convergence is fast and *conservative bound generation* is small. Relaxation of bounds can even be used within clusters because the feedback between sections, while fairly strong, is generally positive.

Relaxation is sometimes used in exact simulation so that analysis can be partitioned among small subcircuits for greater efficiency. Whether this is more efficient than conventional approaches that are modified to specialize in digital circuits depends on the particular application. In a bounding algorithm, the efficiency penalty for considering larger subcircuits is much greater. The relaxation approach also allows a large degree of flexibility in analysis that is essential in the bounding context. The tightness of bounds used can be varied among individual blocks, and the accuracies can potentially be dynamically varied during a relaxation, based on the the worst case uncertainty generated at the previous step. In addition, the occasional blocks that cannot be treated efficiently with a bounding algorithm, such as an analog subcircuit, can be incorporated by using exact analysis on fast and slow input extremes, producing a reasonable and practical substitute for a rigorous bound.

---

[14]The effect of ignoring correlations among device characteristics is classified as *conservative bound specification*. Only the effect of ignoring correlations between two instances of the same device, used for bound generation, is considered in this section.

3.3 Simulating Various Digital MOS Subcircuits

In the previous two sections, the sources of uncertainty that arise from both the specification and the generation of bounds were discussed. The discussion was general because the effects are highly dependent on the circuit being considered. In this section, a range of circuit types found in digital VLSI chips is analyzed in more detail with respect to generating bounds on their behavior using the previously discussed techniques. In each case, the effect of ignoring various correlations is considered, along with the convergence rate for relaxation methods of analysis. Although some of the subcircuits considered are quite rare, they illustrate the important sources of *bounding simplification* in MOS circuit simulation. As shown in this section, some subcircuits are very difficult to analyze efficiently with a bounding algorithm. Such subcircuits can be analyzed with "exact" simulation. Fortunately, the portion of a digital MOS circuit containing these difficult subcircuits is usually small, and the bounding strategy considered here can incorporate exact analysis of some subcircuits in a straightforward manner due to the use of partitioned analysis.

3.3.1 Standard Restoring Logic

The standard restoring logic gate, as pictured in figure 3-12, is the backbone of many digital systems, and therefore must be easily treated by any useful algorithm aimed at digital circuits. A restoring logic gate is constructed from a cluster of transistors divided into pullup and pulldown sections. At least one of these sections provides a relatively low resistance path, except possibly during short transition periods. Coupling between the input and output of a MOS logic gate is primarily in one direction. The internodal capacitance that produces feedback is usually small, making relaxation algorithms converge very quickly.

Restoring logic gates have built-in noise immunity that provides a margin for uncertainty in analysis. If a bounding algorithm does not exceed the noise margin in its d.c. uncertainty, arbitrary uncertainty in voltages during transitions is eventually restored if the inputs are restored. In other words, if a bounding algorithm is reasonable in the steady state, the outputs of a circuit constructed from restoring logic gates will eventually settle some time after the inputs do. There is also an immunity to timing errors in digital circuits due to synchronization by clocks. Hence, uncertainty in delay is restored by clocks while uncertainty in voltage is restored by the circuit.

In a rough sense, the feedback between the inputs and the outputs of a logic

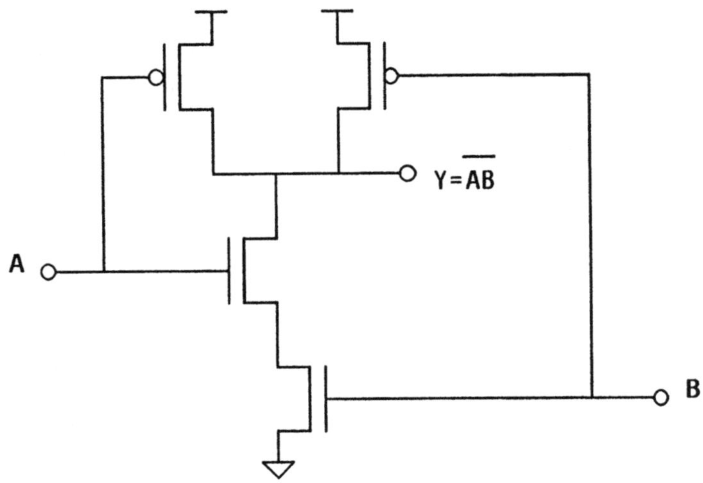

Figure 3-12: Bounding algorithms can analyze restoring logic gates efficiently.

gate appears to be negative. A fast input produces a fast output that slows the input down through the gate's Miller capacitance. The bounding strategy, however, can produce positive feedback in the calculation. An upper bound on an input voltage produces a lower bound on the output voltage, that is then used to calculate an upper bound on the derivative of the output voltage[15]. Since the correlation between a signal and its derivative is not considered, the upper bound on the input is self-consistent, i.e., gives rise to itself through the feedback path. As a result, bound relaxation, when taken to convergence, can avoid additional *bounding simplification* beyond that caused by ignoring correlations between the output voltage and its derivative.

Uncertainty in the analysis of restoring logic gates arises from including unrealistic voltage derivatives. The resulting uncertainty in charge dumped onto each node through coupling capacitors is not large, though, because each

---

[15] As explained further in chapter four, an upper bound on v_B is produced, roughly speaking, from an upper bound on the current entering the load capacitor through $i_L = C_L \dot{v}_B$. An upper bound on i_L can be calculated by assuming v_B is at its lower bound.

node is restored. Any uncertainty in voltage is quickly removed through the low resistance paths to the power supply voltages. What results is a small uncertainty in the effective delay of each logic gate that can be improved by considering bounds on the total charge passing through coupling capacitors.

3.3.2 Pass Logic

Pass transistor logic consists of transistor switches placed between restoring logic gates, as pictured in figure 3-13. If a path through pass logic is enabled, the restored signal at the input is allowed to pass to the output in a possibly degraded form. Logic signals are sometimes sent through pass transistors in both directions, but usually only one direction is used in a given clock cycle. If no paths are enabled to a particular node in a pass network, it becomes a dynamic node. A dynamic node will maintain a logic level for some time because its capacitance is discharged very slowly through leakage currents.

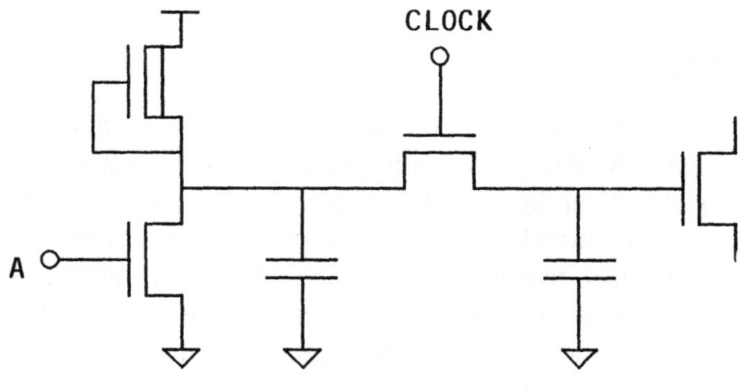

Figure 3-13: Pass networks can contain dynamic storage nodes.

If no dynamic nodes are present in a pass circuit, it can be considered to be a simple extension of a restoring logic circuit. Each pass network is a connection and extension of the transistor clusters associated with the input gates. The feedback between clusters is essentially unchanged from the restoring logic case. Even though pass transistors can be bidirectional, coupling between clusters is still primarily unidirectional. The main difference is that the margin for maintaining the restoring property of the logic is often reduced by uncertainty in pass network calculations. Thus, tighter bounding algorithms are required. In addition, nodes that are not strongly connected to a pullup or pulldown tend to be more sensitive to uncertainty in charge dumped through coupling capacitors. As a result, unrealistic voltage derivatives cause larger delay uncertainties

If a dynamic node is present, another problem can arise that is a serious concern. A dynamic node is extremely sensitive to uncertainty in charge dumped into it through a coupling capacitor, as uncertainty is not continuously restored[16]. Groups of dynamic nodes connected through low resistance paths only to each other have the same problem as a single dynamic node does. Any uncertainty in the group only gets restored when the nodes are refreshed. Even if bounds on leakage current are fairly tight, given enough time they can generate sufficient uncertainty to erase any information stored on a dynamic node. The uncertainty due to ignoring correlations between different times in derivative waveforms can produce large unrestored uncertainty. As a result, if a dynamic node is connected to an internodal coupling capacitor, the total charge passing through the capacitor must be considered to obtain a reasonable bound.

3.3.3 Latches

A latch consists of a feedback loop containing an even number (usually two) of inverting logic gates (transistor clusters) with two stable states, and thus static memory, as pictured in figure 3-14. Since its d.c. solution is not unique, information must be provided for d.c. analysis that indicates which solution is desired. The feedback that is present in the latch circuit also produces slow convergence for waveform relaxation algorithms when the latch is partitioned into its two clusters. Such a partitioning requires that analysis be broken into a number of time windows for efficiency. Exact relaxation algorithms with a smart partitioning routine can combine these clusters into a single block, but a bounding algorithm cannot do this unless the behavior of the group of clusters can be directly bounded.

In exact waveform relaxation algorithms, even if the two tightly coupled clusters of a latch are analyzed separately, relaxation often appears to converge quickly with the right initial guess. If the initial state of the latch is close to one

[16] A precharged logic gate is a fairly common subcircuit that contains a dynamic node, but since it is restored once each clock cycle, it is not overly sensitive to uncertainty in leakage current.

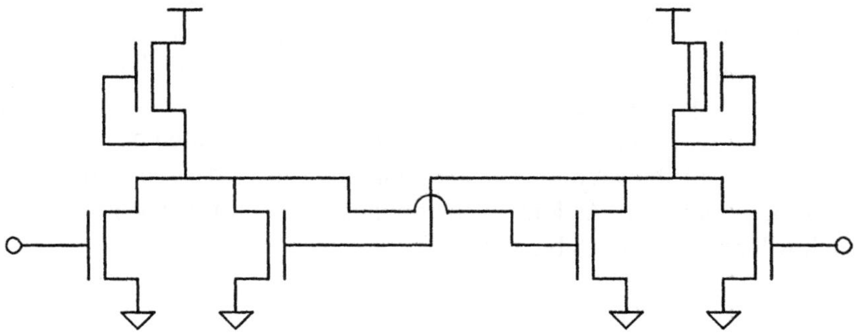

Figure 3-14: The behavior of latches can be tightly bounded.

of the stable equilibria, the simple guess for a behavior where voltages remain at their initial value for all time converges very quickly for the time period before the latch switches state. Fast convergence is really the result of a lucky initial guess that happens to be virtually correct. For different initial guesses, or after latch transitions when the guess is no longer essentially correct, the number of iterations required to reach convergence is proportional to the number of cycles around the feedback loop that a signal could travel during the simulation time period. At each point in time, signals are calculated assuming that there is no feedback, so, after one loop delay, calculations can stray significantly from the solution. As a result, exact waveform relaxation algorithms rarely partition strong feedback loops without dividing analysis into small time windows[17].

A bounding algorithm does not generally have the luxury of being able to start with a good initial guess for the waveform, even when the latch is stable. As discussed further in chapter five, a reasonable initial guess for a bounding algorithm is the interval containing all reasonable behaviors for a circuit, including both good and poor initial guesses for an exact relaxation algorithm. When an initial interval is relaxed, the result must contain all behaviors that result from exact relaxation of both the good and the poor initial behaviors. As a result, if a bounding algorithm treats each cluster in a latch separately, it must break up the analysis of the latch into small time windows.

---

[17] These time windows should have a duration that is roughly equal to the time that it takes a signal to pass once around the feedback loop.

A latch can be bounded just as tightly as the logic gates that it is constructed from, due to the positive nature of its feedback loop. The upper bound on the voltage of any node is calculated with the assumption that the voltage was following its upper bound earlier in time. This self-consistency implies that the bounds are allowable responses, given the bounds used for each logic gate, and the delay accuracy should be similar to that of logic gates. In fact, due to the nature of restoring logic, the behavior of the feedback path often has little effect on the behavior of the circuit other than holding it in its desired state after an input strobe disappears. As long as the bounds maintain this restoring nature, the latch will remain in its stable state.

3.3.4 Ring Oscillators

A ring oscillator consists of a feedback loop with an odd number of inverting logic gates (transistor clusters), as illustrated in figure 3-15, that can be enabled to produce a free running oscillation. The result is a clock signal with a frequency that is related to the speed of the devices on the chip. Such circuits can be partitioned and handled with a relaxation technique, but unless time windows are used, the convergence rate is proportional to the number of oscillation cycles considered, i.e., only one cycle is produced correctly in each iteration.

With a good initial guess, e.g., constant voltage waveforms, an exact relaxation algorithm does not generally need to calculate many cycles of activity in any one iteration, as cycles that have been analyzed a few times have converged and those that have not appeared are latent. In this case a relaxation algorithm can be very efficient in computation even without considering separate time windows. As with latches, though, bounding algorithms cannot count on good initial guesses, and time windowing can become more important.

With respect to bounding algorithms, the ring oscillator does not look promising due to its strong negative feedback. The upper bound for any node voltage is produced by assuming that during the previous cycle, it was at its lower bound. Ignoring the fact that a real waveform could not jump between bounds in this manner produces bounds that diverge from one cycle to the next. Assume, however, that this problem could be solved, and the delay of each gate could be calculated to within roughly ±5% through all cycles. Therefore, the delay around the loop could also be calculated to within about ±5%. After about 5 cycles (10 transitions), the uncertainty in each edge would be about half the period of the oscillator, and the logic state of any node would be completely unknown at all times thereafter.

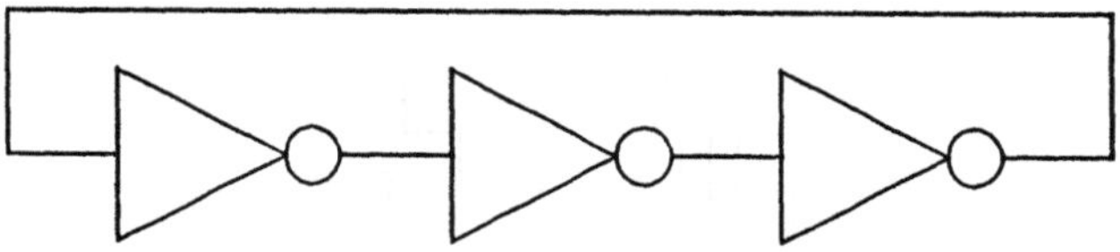

Figure 3-15: Ring oscillators restore signals in voltage but not in time.

The problem with a ring oscillator is not really in the realm of the bounding algorithm used, but rather in the characterization of its behavior. Even an exact simulator cannot predict reliably the voltage of the output after many cycles of operation. The bounds on the behavior merely reflect the uncertainty in the voltage that is produced by the slightest uncertainty in the model or by the slightest simplification made in calculating the response. The output of an oscillator is best understood, not as a particular value of voltage at a particular time, but rather as a frequency that must be within a certain range. A bounding algorithm could conceivably be used on such a circuit with a modification of the definition of circuit behavior, but an exact simulation, producing one reasonable approximation to the behavior is probably more appropriate. Interestingly, this is one situation where it is not desirable to know the uncertainty produced by a simulator in terms of voltage waveforms. With the standard definition and partial ordering for circuit behavior, *conservative bound specification* produces useless results.

3.3.5 Schmitt Triggers

A Schmitt trigger, as illustrated in figure 3-16, is a logic gate that uses hysteresis to improve noise immunity. The hysteresis is often produced by using positive feedback within a single cluster. One node of the cluster is used to drive a pullup connected to another node, as in figure 3-16. The convergence of a relaxation algorithm that analyzes clusters separately is not adversely affected by the feedback, since it takes place inside a single cluster. If the algorithm used to analyze each cluster does not handle the case where clusters can have internal feedback through transistor gates, as the algorithm derived in this book does not, the feedback path must be partitioned, i.e., the nodes in the cluster must be analyzed separately. In this case, some reduction in performance occurs if time windowing is not used.

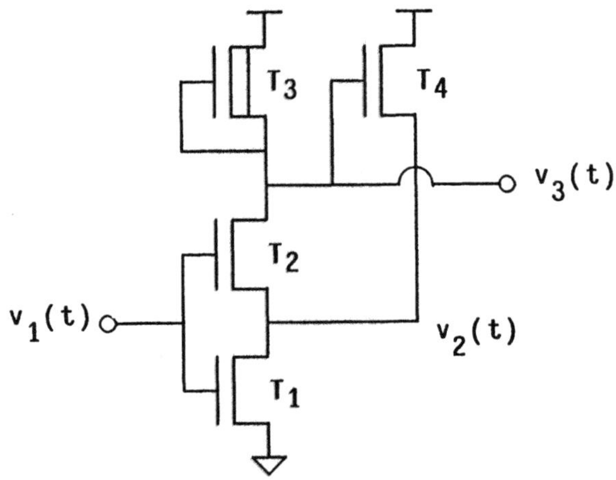

Figure 3-16: Schmitt triggers exhibit positive feedback.

Assuming that the feedback path is cut by partitioning, the bounds that result for the behavior of the entire logic gate after relaxation are good because the feedback is positive. An upper bound on the voltage waveform $v_3(t)$ in figure 3-16 produces an upper bound on the voltage waveform $v_2(t)$ due to the increased pull of transistor T4. The upper bound on $v_2(t)$ also produces an upper bound on $v_3(t)$, because the nodes are coupled through transistor T2. The self-consistent nature of the bounds implies that the extreme behaviors are possible solutions and therefore the coupled bounds are optimal. If the correlation between the gate voltages of T1 and T2 is ignored when generating bounds for each subcircuit, some *bounding simplification* is introduced. However, as long as a bounding algorithm maintains the restoring nature of a Schmitt trigger logic gate, useful bounds on its behavior can be produced.

3.3.6 Bootstrap Drivers

A bootstrap driver contains a large coupling capacitor that produces strong feedback between transistor clusters, as pictured in figure 3-17, and allows a dynamic node to charge much higher than the supply voltage. Due to the bidirectional coupling that results, a relaxation algorithm that treats the two clusters separately does not converge very quickly. Relaxation with simple partitions, i.e., those whose blocks are simply clusters, is only feasible when bootstrap drivers comprise a small fraction of the circuits on a chip. Fortunately for circuit analysis, bootstrap circuits are quite rare.

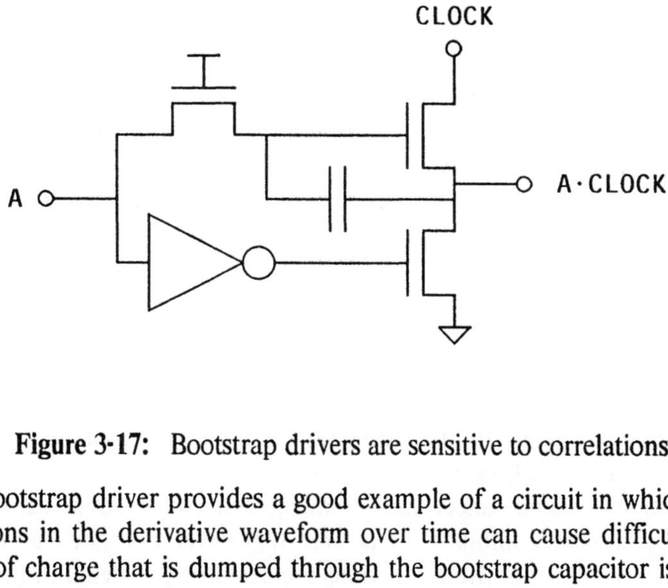

Figure 3-17: Bootstrap drivers are sensitive to correlations.

The bootstrap driver provides a good example of a circuit in which ignoring correlations in the derivative waveform over time can cause difficulties. The amount of charge that is dumped through the bootstrap capacitor is critical to the behavior of the circuit, and using bounds on the voltage derivatives produces very poor estimates for this quantity. Bootstrap circuits use the unusual and difficult combination of a dynamic node with a large coupling capacitor.

In a rough sense, the feedback in a bootstrap driver is positive. A fast waveform on the output couples through the bootstrap capacitor to produce a fast waveform on the dynamic node. This fast dynamic node waveform then produces a fast waveform at the output. The bounding algorithm used for each transistor cluster, though, can easily produce a situation where the feedback in the calculation is negative. Because derivative waveforms are decoupled from their corresponding signals, an upper bound on the derivative is normally calculated from a signal's lower bound. Therefore, the slow rising output waveform produces the largest output derivative, and thus the fastest dynamic node waveform. The combination of charge dumping uncertainty and negative feedback in the calculation makes the generation of behavior bounds that are tight enough to capture the bootstrap operation very difficult.

3.4 Applications for Bounding Algorithms

The application of an "exact" circuit simulation algorithm is fairly straightforward. It can be used to predict the behavior of a circuit when given a set of inputs along with a circuit model. Sometimes parameters such as accuracy thresholds can be adjusted, but they are seldom moved from the default values chosen by experimentation.

More variety exists in the possible application of a bounding algorithm. Bounds can be adjusted in accuracy to perform different functions ranging from "exact" analysis to rough timing simulation. Whole classes of behaviors can be considered simultaneously, as is often done within procedures such as input-independent timing analysis. The accuracy can be dynamically adjusted based on partial simulation results. In addition, the circuit model and excitation can be specified in many different forms.

A bounding algorithm provides two major functions in a circuit simulator. First, a bounding algorithm can be used to simplify computation, i.e., trade computation for measured uncertainty. Second, it can be used to determine the effects of input uncertainties, including uncertainties in the circuit model itself. Input-independent analysis is one example of an application where both of these functions, called uncertainty management and worst case analysis, respectively, can be combined to answer high-level questions about digital circuit behavior. Both of these functions are examined in detail in this section. In addition, input-independent analysis is presented as an illustration of one use of both functions.

3.4.1 Uncertainty Management

Current attempts to manage uncertainty in VLSI circuit simulation are useful but not well developed. A circuit designer does not always require detailed simulation, especially at early stages of a design, and thus often uses very rough circuit models to speed computation. Some tasks such as rough timing verification and fault simulation use simple models as well, because exact waveforms are not essential to the questions being answered. These methods introduce uncertainty into a simulation task that is difficult to measure and adjust in a rational manner.

There is a similar situation for uncertainty management within a single simulation. Mixed-mode simulators attempt to use different levels of accuracy throughout a circuit. A designer can specify a critical path that must be simulated exactly, while other portions of the circuit are treated with switch-level simulators. These methods require extensive intervention by the user, and again, uncertainty is difficult to measure and adjust in a rational manner.

A bounding algorithm can potentially be used to adjust accuracy in a more efficient and automated fashion, and produce a result in which the appropriate level of trust can be placed, i.e., in which the amount of uncertainty that results is bounded. A simple example is provided here to illustrate the idea of uncertainty management using gate models characterized simply by a delay.

The circuit in figure 3-18 is constructed from logic gates that each have a delay of 10 ns. The three inputs arrive at $t=0$. If each logic gate is bounded to within ±100%, v_1 settles within the bounds $0 \leq t \leq 20$ ns, v_2 settles at $0 \leq t \leq 40$ ns, v_3 settles at $0 \leq t \leq 20$ ns, and v_4 settles at $0 \leq t \leq 60$ ns. The output of the circuit v_4, which actually settles at $t=30$ ns, is also bounded to within ±100% uncertainty.

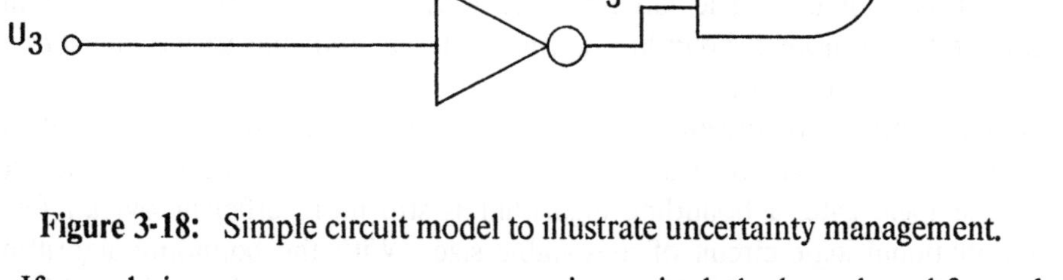

Figure 3-18: Simple circuit model to illustrate uncertainty management.

If, as a design progresses, more accuracy is required, the bound used for each gate can be adjusted to one with ±10% uncertainty. The result is then that v_1

settles within the interval [9,11], v_2 settles within [18,22], v_3 settles within [9,11], and v_4 settles within [27,33], all in units of ns. The uncertainty in the output also falls to ±10%. If a relaxation method were used on a more complex model, one could easily imagine stopping the iteration when a desired level of accuracy is achieved. The accuracy of the bounds on blocks could be gradually increased as relaxation continued.

In the example just described, it is clear that the analysis of the logic gate producing the signal v_3 is not as critical as the others. Since it is not in the critical path, it need not be upgraded from ±100% analysis to ±10% analysis to cause the output to be upgraded from ±100% to ±10% uncertainty. If a relaxation algorithm were being used, or just a series of experiments in the case of the simple circuit model, a smart simulation algorithm could concentrate on paths that are more likely to be in the critical path when trading computation for accuracy. A good strategy would decrease the amount of computation required to produce a given level of accuracy in the circuit outputs.

A particular bounding algorithm is only useful for uncertainty management if it requires significantly less computation than an "exact" algorithm. Comparisons with algorithms that produce rough approximations, though, is more difficult. Bounded uncertainty is more valuable than unbounded uncertainty, but its extra value in terms of computation is application dependent and difficult to quantify. The following example provides a simplified comparison.

If a combinational logic block were being designed that had to be faster than 10ns, a simulator that estimated a 9ns delay for its critical path would not be helpful unless its standard deviation were only a few percent. If the number of paths with a delay close to 9ns were large, even more statistical accuracy would be required to achieve high confidence that the specification was met. If bounding algorithms were used, only enough computation would be required to achieve a ten percent level of accuracy for the critical paths, regardless of the number considered. The larger a circuit is and the more cycles that are simulated, the more a bounding approach is favored due to the statistics of approximate simulation.

Even if bounded uncertainty were an order of magnitude more costly to calculate than approximate uncertainty measured at one standard deviation for a single logic gate, a bounding algorithm is still more efficient overall for a combinational logic circuit of reasonable size. With the bounding algorithm, uncertainty can be managed to concentrate on critical paths and confidence does not decrease with circuit size. An efficient bounding algorithm that allows a wide range of computation levels and a corresponding range of accuracy levels would be a useful tool in a VLSI simulation package.

3.4.2 Worst Case Analysis

A circuit designer is often concerned with the performance of a circuit over a fairly wide range of inputs and circuit models. The uncertainties in the simulation, often arising from variations in fabrication processes, produce a range of circuit behaviors that can be captured in a bound. Since complete bounding simulators have not yet been available to circuit designers, they often use a technique called approximate worst case analysis to estimate the range of behaviors. In approximate worst case analysis, an exact simulator is used at least twice, with inputs that are at the extremes thought most likely to produce extreme behavior. One use of a bounding simulator is to replace this approximate procedure with a rigorous one.

If tight bounds on the effect of uncertainty in the excitation or the model are generated, a bounding algorithm produces essentially an exact analysis of an appropriate model with bounding behavior. The main source of *bounding simplification* is conservative specification of the bounds on the excitations and models. As mentioned in the first section of this chapter, the most significant source of uncertainty amplification associated with this simplification comes from ignored correlations among different devices in the circuit. As a result, the bounding strategy proposed here is quite conservative when incorporating uncertainties in devices, but can incorporate input uncertainties efficiently. The following examples illustrate the main ideas of worst case analysis.

Consider a very simple model for a digital circuit in which all signals are step functions and all logic gates are unidirectional, with a certain amount of internal delay. Assume that a logic gate is being simulated that has a delay expression given by $D=1+A+B$, where A and B are internal gate variables associated with different devices. The delay is monotonic in both A and B. Now consider an internal parameter x that is known to range between 0 and 2. If $A=\min(4x,4)$ and $B=2-x$, as pictured in figure 3-19, the delay has a maximum possible value of 6 over the possible range of values for x. If the total delay equation could not be solved for x in closed form, in general the solution would have to be obtained for all possible values of x and the maximum used. The delay is a monotonically increasing function of A and B but A is an increasing function of x while B is a decreasing function of x. Therefore, the delay is not monotonic in x and calculating the delay for $x=0$ and $x=2$, as would be done in approximate worst case analysis, is not sufficient to bound the delay.

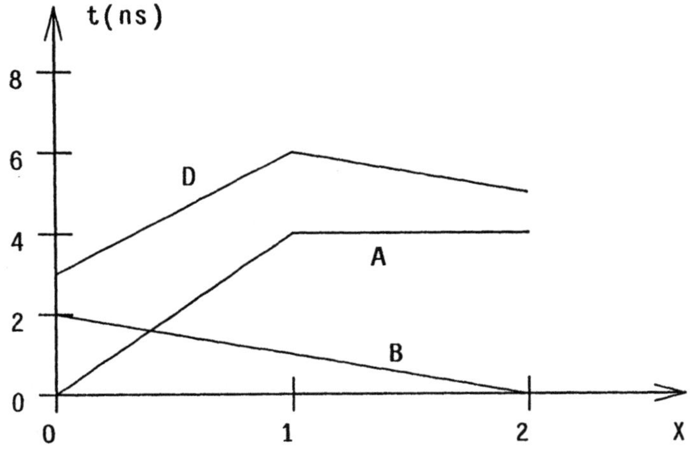

Figure 3-19: Simple example of a circuit model containing uncertainty.

Now consider applying bounding techniques to the previous problem. The variables A and B are part of the specification for the circuit model so their bounds must be specified. Since A is an increasing function of x, its maximum value of 4 is obtained with the largest possible value of x. In a similar manner the maximum value of B can be calculated[18] as 2. Now, recalling that D is monotonic in both A and B, using their maximum values in the calculation of D produces an upper bound of 7. The correlation between A and B is ignored in the model specification, producing a conservative estimate for the behavior. The search space of possible delays has been reduced to simple bounding "corner" calculations at the expense of some accuracy.

Even if the model is completely specified, say $A=3$ and $B=1$, the input to the logic gate might be unknown. Assume that the inputs to the logic gate, step functions in the simple model, are guaranteed to switch between $t=0$ and $t=1$.

---

[18] Even if A and B were not monotonic in x, they could be bounded by trying many values for x, as they represent a device model that need only be evaluated once for each new fabrication process. More generally, correlations among variables inside a device model can be easily incorporated because computation is not a limiting resource in bounding device models, only in bounding circuit behaviors.

Since the delay of the logic gate is known to be 5, the output must switch between t=5 and t=6. In this case there is no *bounding simplification.*

As a more realistic example of worst case analysis consider an uncertainty in gate oxide thickness for an NMOS process. A decrease in oxide thickness increases the current functions of both the pullup and pulldown transistors and can increase or decrease their d.c. output voltage. Such a decrease is not even guaranteed to speed up a circuit because the load capacitances will also rise. As discussed in the next chapter, to generate an upper bound on the response of a single logic gate, the maximum conductance is used for the pullup and the minimum conductance is used for the pulldown. As a result, the minimum value of oxide thickness must be used when calculating the current function of the pullup and the maximum value used when calculating the current function of the pulldown. To generate a lower bound on the response, the maximum value of oxide thickness must be used for the pullup and the minimum value used for the pulldown. In either case, the load capacitance used for each logic gate depends on whether its output is rising or falling. Any circuit behavior bound will generally be calculated using both extreme values for oxide thickness throughout the entire circuit. Since every device in a VLSI circuit generally has approximately the same gate oxide thickness, the bound produced can be quite conservative. The approximate approach uses the largest value of oxide thickness to generate the slowest circuit, because speed tends to be a monotonic function of oxide thickness for standard circuit forms.

Worst case analysis is a potential application for a bounding simulator, but if large uncertainties in devices are considered, further theoretical work is needed to achieve tight bounds. For this purpose the circuit model must be analyzed at the level of parameters such as gate oxide thickness, not at the level of device characteristics. The analysis done in this book is aimed primarily at uncertainty management and worst case analysis when applied to inputs. Approximate worst case analysis, when applied to model uncertainty, is adequate for many circuits.

3.4.3 Input-Independent Analysis

One common problem in digital circuit analysis that can combine both uncertainty management and worst case analysis is input-independent analysis. Often, for timing verification of a combinational logic circuit, one desires to find the range of possible outputs when all possible input waveforms are considered. The inputs are usually restricted to a particular class such as those that have settled to beyond either logic threshold before a particular time. Even if each of n inputs to a circuit could only be one of four possible waveforms: rising, falling, constant high, or constant low, 4^n separate exact simulations are required to determine the exact range of possible outputs. For even fairly small circuits, the exponential growth of computation makes exact simulation infeasible.

The conventional approach to solve this problem is to use rough bounding techniques. Conservative estimates of switching time are used for each gate. To calculate an upper "bound" on the delay, each gate is only allowed to switch after its last input has arrived. Since only one quantity is calculated for each gate, namely a bound on its switching time, computation scales linearly with circuit size. With a sophisticated bounding algorithm, this process can be made rigorous, and tighter bounds can be achieved without losing the linear scaling.

Worst case analysis arises in the specification of the circuit excitation. Since each input can have four distinct types of waveforms, each with a range of possible characteristics, a simple interval of excitations using the partial ordering defined in section 3.1 is inadequate to capture the information in the excitation subset. The tightest interval bound would contain all waveforms with voltages between ground and the supply voltage. This problem can be avoided by using a union of waveform intervals, as pictured in figure 3-20. The first graph in the figure represents two waveform intervals for a single input voltage source. Any valid waveform must lie completely between the two rising bounds, or lie completely between the two constant bounds. Only two of the four waveform types are pictured for simplicity. If the input is specified as a union of intervals, each interval can be processed separately by a bounding algorithm and the union of the resulting behavior intervals is a valid bound on the behavior.

Worst case analysis alone is sufficient to determine the range of possible outputs rigorously and generate tight bounds that take into account details about waveform shapes. However, the exponential computation costs of separately analyzing each case still exist. Uncertainty management can be used to reduce the computation to the point where it scales linearly with circuit size without sacrificing information about waveform shapes. The key to achieving linear scaling of computation with circuit size is to keep the amount of information needed to characterize the output of each gate the same throughout the circuit. If each output is characterized by only four waveform intervals similar to those used to characterize the input voltage sources, information about waveform shapes can be maintained without requiring any growth in the number of waveform intervals considered. To compress the characterization of any waveform to only four intervals, one need only take all intervals of the same type, i.e., rising or falling[19], and consider only their union. All possible rising waveforms at a gate output, for example, can be restricted within (envelope) bounds that apply in all cases where the output rises.

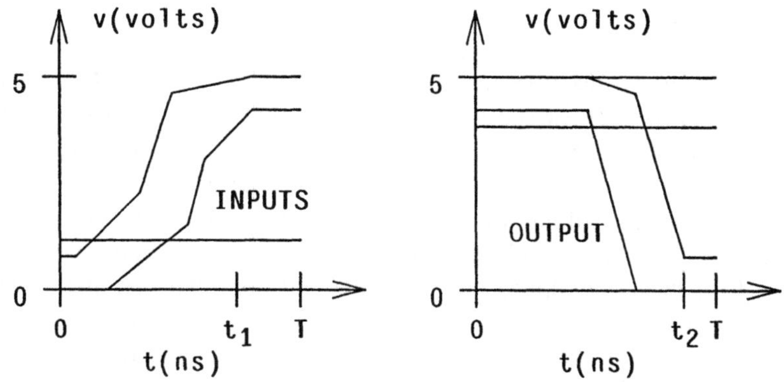

Figure 3-20: Bounds shrink the search space in input-independent analysis.

As an example, consider a 3-input NOR gate. Assume that each input might be either a rising or a constant low waveform as pictured in the first graph of figure 3-20. The graph represents only one of the inputs, which is guaranteed to settle by time t_1, but the others are assumed to be similar. If each combination of input intervals is used to compute the possible gate outputs, eight output intervals are produced, one of which is constant high and the other seven of which are falling. If a single falling output bound is generated from the minimum of the seven falling lower bounds and the maximum of the seven falling upper bounds for all $t \in [0,T]$, then the eight cases are compressed into two, as pictured in the second graph of figure 3-20. If the output is a constant high, one set of bounds applies and if the output is falling, the other set applies regardless of which of the seven sets of inputs generated it.

A bounding algorithm can expand the function of a simulator into the realm of what is sometimes called circuit analysis. Simulation often denotes predicting

[19]Waveforms that first rise and then fall are possible in combinational logic circuits but are not considered here for simplicity. They can be incorporated with the decomposition pictured in figure 3-5.

the behavior of a fixed circuit with fixed inputs, and analysis often denotes deducing general properties of a circuit that are somewhat independent of specific models or inputs. Bounding algorithms form a bridge between these two tasks.

3.5 Summary

The two main points presented in this chapter are that, first, a bounding approach is feasible for the simulation of large digital MOS circuits, and second, a bounding approach is very useful. Extreme behaviors of a digital circuit correspond roughly to slow and fast responses, and ignoring correlations among signals and subcircuits does not produce overly pessimistic conclusions. When approached at the level of voltage waveforms, digital circuits provide a friendly problem for a bounding approach. A strategy for the definition and generation of behavior bounds that are efficient for most of a typical digital MOS circuit was presented. In addition, it was shown that bounding algorithms can increase overall efficiency of simulation, a pressing need, by enabling uncertainty to be managed.

4. THE VLSI CIRCUIT MODEL

The last chapter presented a high-level bounding strategy for digital MOS circuits that assumed the circuit model being analyzed had a number of special properties. A general class of networks that exhibits these properties is derived here. Restrictions are placed on the model that insure both the existence of sufficient monotonic relationships and the ability to partition analysis, but do not significantly restrict the model's generality.

Many hypotheses about the monotonic properties of MOS circuit models are intuitively appealing. The current charging any grounded capacitor should be underestimated if its voltage is overestimated. The output drive of any logic gate should be a monotonic function of the individual drive capabilities of each of its transistors. Fortunately, many convenient monotonic relationships such as the two above are largely true and hold for a wide range of common models and situations. This chapter discusses both their limits of applicability and relevance to bounding algorithms in detail.

To partition the bounding analysis of a MOS circuit, an interval extension of Waveform Relaxation can be used. The details are postponed until the next chapter, but the extension is based on the convergence of an underlying exact relaxation algorithm. Here the restrictions that must be placed on the circuit model to guarantee this convergence are investigated.

This chapter develops the VLSI circuit model in a number of steps. The device models are defined first. Then the effect of each element type is considered separately. Finally the interaction of the elements is investigated. Roughly speaking, the circuit model can be any lumped element network containing nonlinear capacitors, nonlinear resistors, and resistive (memoryless) transistor models, all with physically reasonable monotonicity and continuity properties. Equations are formulated based on nodal analysis, and both charge and voltage are considered for state variables. While charge is not used as a state variable in the algorithms presented later, it is potentially useful in some circuits as mentioned in chapter three.

4.1 Elements in the VLSI Circuit Model

A wide range of circuit models is commonly used for MOS integrated circuits. Most of these consist of a network of lumped elements, including capacitors, resistors, and (purely resistive) n-channel and p-channel transistors. Precise definitions for these elements that are compatible with most models, along with criteria for comparing two elements of the same type, are presented below. The elements are used in following sections to construct a lumped-element VLSI circuit model.

4.1.1 Resistors

All resistance in a MOS integrated circuit outside of a transistor is modeled here with lumped, two-terminal resistors. Wire resistance can be modeled by a resistor that is close to linear while diffusion to substrate resistance can modeled by a resistor with a diode characteristic. All resistors appearing in the model must be R-elements, that satisfy the mild constraints[20] given in the following definition.

> **Definition 4-1**: An R-element is a 2-terminal resistor that is passive (i.e., $v \cdot i \geq 0$), strictly monotone (i.e., for any two distinct operating points, $\Delta v \cdot \Delta i > 0$), and characterized by both $v = f(i)\ \forall i \in \mathbb{R}$ and $i = g(v)\ \forall v \in \mathbb{R}$, where f and g satisfy a global Lipschitz condition.

A partial ordering for resistors is defined based on current as a function of voltage. This ordering is referred to as a simple ordering when it is being distinguished from others. A resistor is said to be larger than another if the

---

[20] There are a number of relatively obscure constraints that appear in the model definitions, such as the Lipschitz condition, that are only of technical interest and are not generally a concern in practice.

magnitude of its current is smaller for all voltages. Using this ordering, a resistor that is close to linear can often be tightly bounded by linear ones as shown in figure 4-1. The simple partial ordering is consistent with the familiar special case of linear resistors, where the nominal "value" of the resistor in ohms is used to produce a linear ordering. The ordering guarantees that, even for nonlinear resistors, the resistor created by connecting two resistors in parallel is never larger than either of its components. Likewise, a resistor formed by connecting two others in series is never smaller than either of its components.

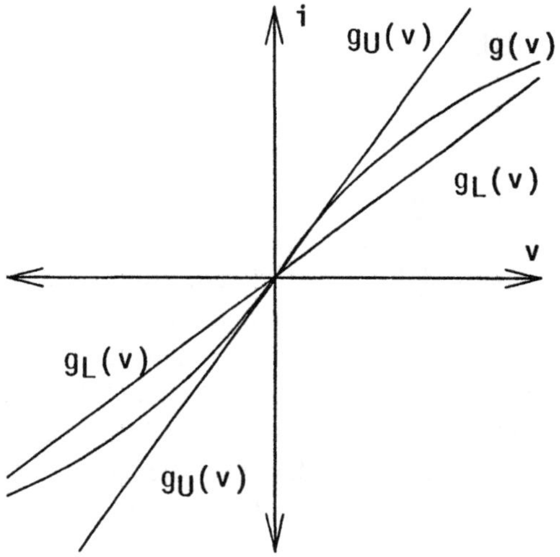

Figure 4-1: Nonlinear resistors can be bounded with linear ones.

Another partial ordering for resistors is also defined, called a current-wise ordering. In this case, the current is considered without taking its magnitude. As a result, the assignment of reference direction for a resistor affects the polarity of this ordering. Since all resistor i-v curves must pass through the origin, the simplest bounds produced by this ordering tend to be piecewise-linear with different segments in the first and third quadrants. Simple bounds and current-wise bounds can be derived from each other simply by exchanging upper and lower bounds in the first quadrant.

Definition 4-2: Given two R-elements R_1 and R_2 characterized by $i=g_1(v)$ and $i=g_2(v)$ respectively, we say that "$R_1 \geq R_2$ (simply)" iff $v[g_2(v)-g_1(v)]\geq 0 \ \forall v$ (equivalently, iff $i[f_1(i)-f_2(i)]\geq 0 \ \forall i$). We say that "$R_1 \geq R_2$ current-wise" iff $g_1(v)\geq g_2(v) \ \forall v$.

The restrictions placed on the resistor models in definition 4-1 do not preclude any reasonable resistor models used in practice. Resistors formed by integrated circuit wires and standard semiconductor junctions all meet the passivity and incremental passivity constraints. Difficulty with the model would only be encountered if devices such as tunnel diodes were present in the MOS circuit being considered.

4.1.2 Transistors

Transistors in MOS integrated circuits are modeled with lumped, multi-terminal resistive, i.e., memoryless, elements. Both the n-channel and p-channel models pictured in figure 4-2 are considered. Any capacitance in a physical transistor is modeled with external lumped (possibly nonlinear) elements as shown in figure 4-3. The transistor models contain all the important monotonic properties of MOS transistors and are general enough to include even the most sophisticated d.c. MOS transistor models, e.g., those including short channel and body effects. Transistor models that meet the required constraints, as listed in the following definition, are called N-elements and P-elements.

Definition 4-3: An N-element or P-element is a 4-terminal resistive element as pictured in figure 4-2 where $i_G=i_B=0$ and $i_D=-i_S=F(v_G,v_D,v_S,v_B)$. The function[21] $F:\mathbb{R}^4\rightarrow\mathbb{R}$ is globally Lipschitz, F evaluates to zero if $v_D=v_S$, and F is a strictly

---

[21] The function can be left undefined in regions in which the diodes between the substrate and the source and drain terminals are not reversed biased. This restriction can then be carried through the following definitions and theorems. For simplicity, it is assumed here that the model is extrapolated into this region and external diodes can model injection of current into the substrate.

monotonically increasing function[22] of v_D and a strictly monotonically decreasing function of v_S. $|F|$ is a monotonically increasing function of v_G for an *N*-element and a monotonically decreasing function of v_G for a *P*-element. $|F|$ is a monotonically decreasing function of v_B for an *N*-element and a monotonically increasing function of v_B for a *P*-element. Furthermore, if $v_G = v_D$ or $v_G = v_S$ (a "diode" configuration), then $G(v_D, v_S, v_B) = F(v_D, v_D, v_S, v_B)$ or $F(v_S, v_D, v_S, v_B)$ is a strictly monotonically increasing function of v_D and a strictly monotonically decreasing function of v_S in either case.

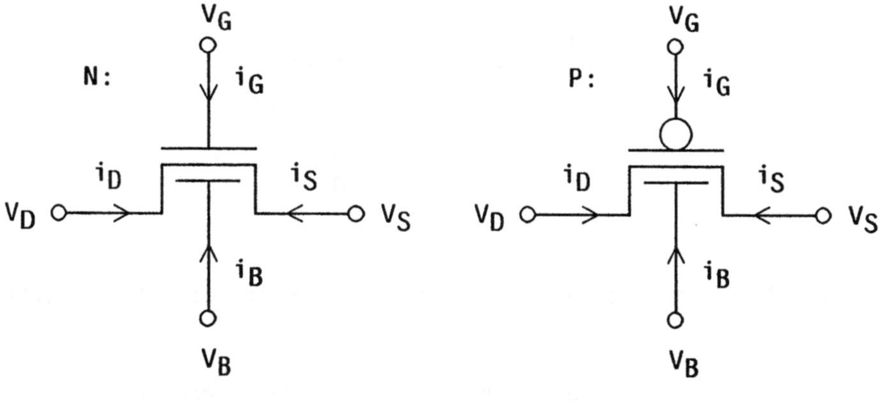

Figure 4-2: *N*-elements and *P*-elements are four terminal resistive elements.

A simple partial ordering for transistors is defined using the magnitude of drain current. Thus, an upper bound on a transistor model could be that of a similar transistor of larger width. The partial ordering is a generalization (with a change in sign) of the one used for two-terminal resistors if a MOS transistor is

---

[22]Models that assume constant drain current as a function of v_{DS} in saturation are not strictly monotonic as assumed here, but they can be made to be so with an arbitrarily small change. Strict monotonicity is required only to insure the existence of a unique solution when two transistor models are placed in series.

viewed as a controllable resistor extending from its source to its drain. A current-wise ordering is also defined that is a generalization of the current-wise ordering for resistors.

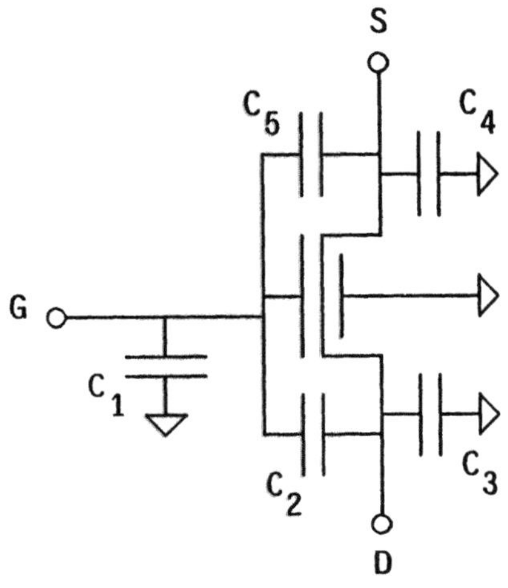

Figure 4-3: A physical transistor is modeled using external capacitor elements.

Definition 4-4: Given two 4-terminal *N*-elements (or two *P*-elements) T_1 and T_2, we say that "$T_1 \geq T_2$ (simply)" iff $[F_1(v_G,v_D,v_S,v_B) - F_2(v_G,v_D,v_S,v_B)](v_D - v_S) \geq 0 \;\forall\; v_G, v_D, v_S$, and v_B. We say that "$T_1 \geq T_2$ current-wise" iff $F_1(v_G,v_D,v_S,v_B) \geq F_2(v_G,v_D,v_S,v_B) \;\forall\; v_G, v_D, v_S$, and v_B.

The restrictions placed on the transistor models here allow most reasonable MOS transistor models used in practice. One is unlikely to find a physical MOS transistor that violates the d.c. monotonicity constraints. Only models that require distributed capacitance within the transistor are not allowed. As shown in section 4.6, such models do not have simple monotonic properties, and hence are more difficult to treat efficiently. Fortunately, fairly sophisticated models can be constructed with only external lumped capacitance.

4.1.3 Capacitors

All capacitance in a MOS integrated circuit is modeled with lumped, two-terminal capacitors. Nonlinear capacitors are often required in a good model so they are included along with linear ones. Capacitor elements model the capacitance between terminals of a transistor as well as the capacitance between signal wires, including ground. All capacitors appearing in the model must be *C*-elements that satisfy the mild constraints in the following definition.

> **Definition 4-5**: A *C*-element is a 2-terminal capacitor characterized by $v=f(q)\ \forall q\in\mathbb{R}$ and $q=h(v)\ \forall v\in\mathbb{R}$ where f and h are continuously differentiable functions passing through the origin. The incremental capacitance $c(v) = h'(v)$ is Lipschitz in v and bounded from above and below, i.e., there exist real numbers $\alpha, \beta > 0$ such that $\alpha < c(v) < \beta\ \forall v$ (this implies that f and h are Lipschitz).

A simple partial ordering for capacitors is defined using the magnitude of the charge. This partial ordering is analogous (with a change in sign) to that used for resistors. A charge-wise partial ordering is also defined, in which the charge is used along with its sign.

Another partial ordering for capacitors, not analogous to either ordering for resistors, is defined based on the incremental capacitance. A capacitor is said to be incrementally larger than another if its incremental capacitance is larger for all voltages. Since the incremental capacitance of each capacitor is bounded, an incremental bound on a capacitor always exists that has constant incremental capacitance, i.e., is linear, as shown in figure 4-4. The nonlinear capacitor pictured in figure 4-4 has the general form of a diffusion to substrate capacitance found at the drain and source of a transistor. For linear capacitors the incremental partial ordering is equivalent to the simple partial ordering. The choice of which partial ordering to use is a function of the choice of state variable, as is discussed further[23] in section 4.4.

---

[23] Also presented later is a fourth partial ordering that is analogous to the charge-wise ordering but applies to incremental analysis. A generalization of incremental analysis must be made before the fourth ordering can be defined.

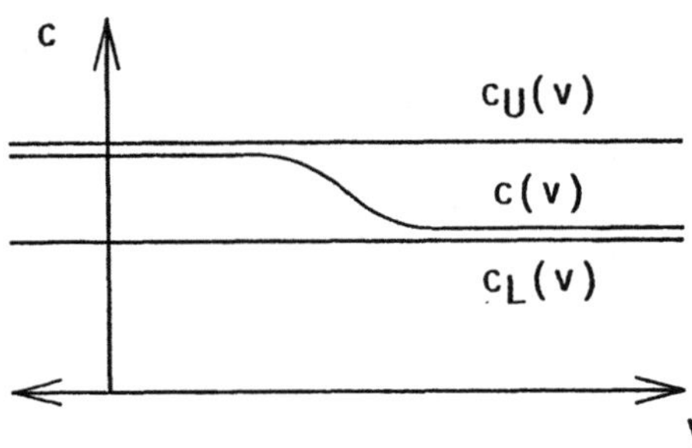

Figure 4-4: Incremental bounds on capacitors limit incremental capacitance.

Definition 4-6: Given two *C*-elements C_1 and C_2 characterized by $q=h_1(v)$ and $q=h_2(v)$ respectively, we say that "$C_1 \geq C_2$ (simply)" iff $v[h_1(v)-h_2(v)] \geq 0$ $\forall v$. We say that "$C_1 \geq C_2$ charge-wise" iff $h_1(v) \geq h_2(v)$ $\forall v$. We say that "$C_1 \geq C_2$ incrementally" iff $h'_1(v) \geq h'_2(v)$, i.e., $c_1(v) \geq c_2(v)$, $\forall v$.

The capacitor model presented here allows most reasonable capacitor models used in practice[24]. The monotonicity constraint is exhibited by real devices found in MOS integrated circuits. One limitation is that the three-terminal capacitor, used in some special circuits, cannot be easily modeled with two-terminal elements. Also, pure distributed capacitance such as that found along a wire cannot be included in the lumped element model. However, distributed RC lines can be modeled very well by a lumped network that contains a sufficiently large number of elements.

[24]Recently proposed nonreciprocal multi-port capacitor models [40] cannot be constructed from two-terminal capacitor elements.

4.1.4 Element Composition Rules

A model for a MOS integrated circuit consists of a network $\mathcal{N}$ that is constructed from the basic elements just described. At this point the only constraint on network $\mathcal{N}$ is that it must contain only *R, N, P,* and *C* elements as defined here. As mentioned in chapter three, all input nodes are connected to grounded input voltage sources. In the following sections, further constraints are imposed that, while having little impact on the usefulness or generality of the model, guarantee general properties that are used to generate efficient bounds on the network's response.

For further analysis, the network is divided as pictured in figure 4-5. All input nodes (including ground) are extracted and labelled 0 to $m-1$. By convention node 0 is reserved for ground and node 1 is reserved for VDD, the power supply input. All remaining internal nodes connected to at least two of the four different types of elements (*R, N, P,* and *C*) are extracted and labelled m to $n-1$. The network $\mathcal{N}$ is then split up into four subnetworks[25] $\mathcal{N}_R$, $\mathcal{N}_N$, $\mathcal{N}_P$, and $\mathcal{N}_C$, each containing only the element type corresponding to its name.

The four subnetworks, each consisting of only one element type, are considered separately to simplify analysis. The subnetworks only interact through the internal nodes m to $n-1$, called "independent nodes," so only the terminal constraints of each subnetwork are of concern in determining their interaction. For simplicity it is assumed that the interaction through these "independent nodes" characterizes the behavior of the entire network. These nodes are the ones used in definition 3-5 to specify a circuit's behavior. In other words, it is not essential to know the internal behavior of any of the subnetworks. In practice the desired output of a simulation usually consists of the voltages at some of the "independent nodes," including all nodes connected both to a capacitor and some resistive device. If desired, the internal behavior of a subnetwork can usually be easily calculated from its terminal behavior.

In a very sophisticated MOS circuit model, the four subnetworks would not normally have any internal nodes. Every node in a MOS circuit is generally connected to both a capacitor and a source of current. In this case the four subnetworks become very simple to analyze and the main issue is their interaction. For a model of intermediate complexity, though, capacitance is

---

[25] A subnetwork need not have elements connected to each of its n terminals, i.e., some of its terminals might not be connected to any internal elements.

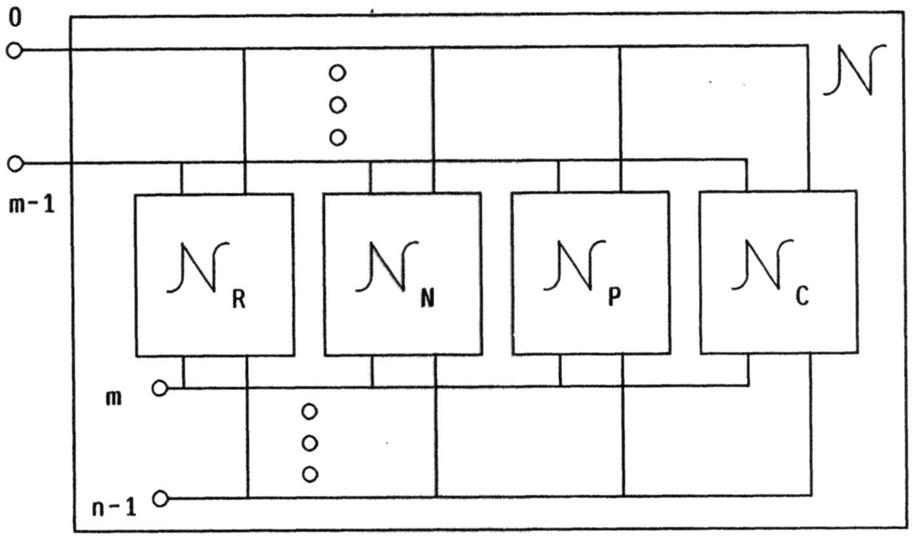

Figure 4-5: The contribution of each element type is analyzed separately.

often ignored on nodes with small area, and thus, small capacitance[26]. In this simpler case the subnetworks actually become more complex. As a result, to allow complex subnetworks, a minimum of constraints on the composition rules for each subnetwork is desired.

The circuit pictured in figure 4-6 is an example of a model of intermediate complexity for a logic gate. The partitioning into element subnetworks is illustrated with dotted lines and one possible set of node labels is given. In this circuit both of the subnetworks $\mathcal{N}_N$ and $\mathcal{N}_R$ have an internal node. In a more sophisticated model of this circuit with capacitance included inside the pulldown network, $\mathcal{N}_N$ would not have such a node. Note that the internal nodes in $\mathcal{N}_N$ and $\mathcal{N}_R$ can be removed simply by connecting them to arbitrarily small grounded capacitors at the expense of increasing overall circuit complexity. Similar tricks can be used, when necessary, to circumvent any of the subnetwork composition rules in practice.

---

[26]Instead of being ignored, this capacitance might also be grouped with others and placed at only one node out of the group.

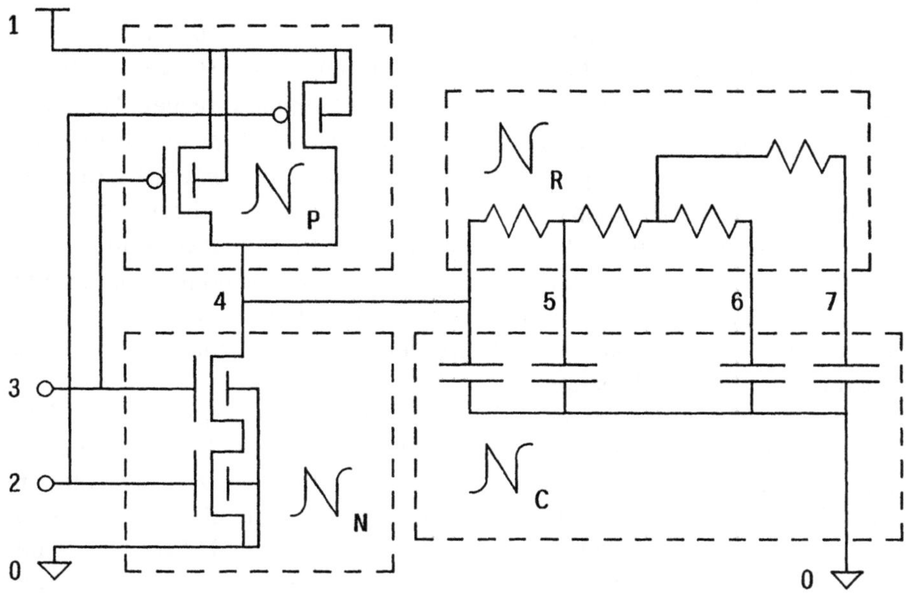

Figure 4-6: A CMOS NAND gate partitioned into four element subnetworks.

Mild constraints guarantee that each subnetwork has simple monotonic relationships between its terminal variables. Only in the presence of such relationships can bounds on behaviors of portions of a network, e.g., the voltages of some of the nodes m to $n-1$, be readily used to calculate bounds on the behaviors of other portions. In addition, mild constraints guarantee simple monotonic relationships between the internal elements of each subnetwork and the subnetwork's terminal constraints. Only when such relationships exist can bounds on element characteristics, including simplifications, be readily incorporated into bounds on circuit behavior. The next three sections show that reasonable composition rules can guarantee monotonic relationships, both among terminal variables and between terminal constraints and element characteristics, that are simple enough to use in computationally efficient bounding algorithms.

4.2 The Resistor Subnetwork

This section considers the subnetwork $\mathcal{N}_R$ of the MOS circuit model that contains all two-terminal resistor elements. It investigates composition rules for $\mathcal{N}_R$ that simplify the generation of bounding information. No restrictions are needed to guarantee monotonic constraints on the terminal variables. However, to guarantee monotonicity with respect to the internal elements, all resistors must be grouped into one-ports that directly connect two external terminals. Within each of these groups, the nonlinear resistors and the incompletely specified resistors, i.e., those that are characterized with resistance bounds, must form a series-parallel subgroup[27]. While these restrictions do not significantly impact the generality of the basic model, they insure that the terminal constraints of the resistor subnetwork are (suitably defined) monotonic functions of the internal resistors.

There are three reasons, beyond the need to incorporate resistors into a MOS circuit model, to consider resistor subnetworks in great detail and to achieve a large degree of generality. First, resistors are often used extensively in simplified MOS circuit models. If all transistors in a circuit are modeled by simple resistors, fairly complex resistor circuits can be created. Second, resistor networks provide a starting point for the analysis of the more complex transistor subnetworks $\mathcal{N}_N$ and $\mathcal{N}_P$. Third, all results for resistor networks lead to some analogous and potentially useful properties of the capacitor subnetwork $\mathcal{N}_C$.

4.2.1 Terminal Constraints

The resistor subnetwork $\mathcal{N}_R$, as pictured in figure 4-7, affects the rest of the circuit through its terminal constraint $\mathbf{i}=\mathbf{f}^R(\mathbf{v})$. In this notation $\mathbf{i}$ and $\mathbf{v}$ represent vectors whose components are the terminal currents and voltages respectively. The incremental passivity constraints placed on the resistors in section 4.1 are sufficient to guarantee a unique solution for the currents and voltages in $\mathcal{N}_R$ given the terminal voltages, and therefore they guarantee a well defined mapping $\mathbf{i}=\mathbf{f}^R(\mathbf{v})$. The currents are expressed as a function of the node voltages here for a number of reasons. First, some of the terminals of $\mathcal{N}_R$ can be

---

[27]This restriction does not rule out arbitrary one-port resistor groups if the resistors are linear.

connected to fixed voltage sources[28], preventing the terminal voltages from being exclusively dependent variables. Second, the primarily local connections found in MOS circuits simplify the calculation of terminal currents as a function of terminal voltages. Finally, in a circuit without inductance, currents are used to derive new states, as time progresses, from old states characterized by voltages. As a result, calculations naturally consider the terminal currents to be the dependent variables. The choice of voltages as independent variables foreshadows the use of nodal analysis, discussed further in section 4.4. Nodal analysis typically produces a very efficient formulation of the network equations for integrated circuit models.

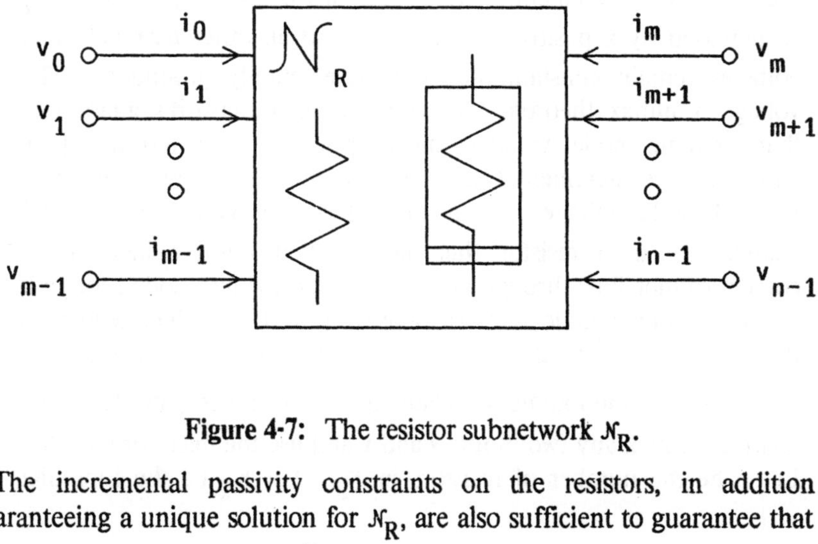

Figure 4-7: The resistor subnetwork $\mathcal{N}_R$.

The incremental passivity constraints on the resistors, in addition to guaranteeing a unique solution for $\mathcal{N}_R$, are also sufficient to guarantee that the terminal constraint function f^R is monotonic. If any terminal node voltage is overestimated (or correct) in the evaluation of $f^R(v)$, the corresponding terminal current is overestimated (or correct) and all other terminal currents are underestimated. This intuitively appealing result is stated here as part of a theorem containing the important properties of $\mathcal{N}_R$ that do not depend on topology constraints.

[28]Recall that current sources are not used in the circuit model so there is no corresponding problem when using terminal currents as dependent variables.

Theorem 4-1: Let $i = f^R(v)$ where $f^R: \mathbb{R}^n \rightarrow \mathbb{R}^n$ is defined by a resistor subnetwork $\mathcal{N}_R$ with any topology. For all integers j and k such that $0 \leq j,k \leq n-1$, i_j is a monotonically increasing[29] function of v_k for $k=j$ and a monotonically decreasing function of v_k $\forall$ $k \neq j$. Furthermore, f^R satisfies a global Lipschitz condition.

Proof: Existence and uniqueness[30] of $f^R(v)$ is shown in [41]. To examine monotonicity, pick any node k for $0 \leq k \leq n-1$ and consider the network in incremental terms about any operating point when v_k is increased by a positive amount Δv_k and all other terminal node voltages remain constant, i.e., are incrementally grounded. The voltage minimax theorem for passive resistor networks guarantees that extreme node voltages must appear at the terminals [42]. Applied in an incremental sense, it guarantees that any incremental internal node voltage lies in the interval $[0, \Delta v_k]$. Incremental passivity of each resistor then implies that positive incremental current cannot flow through any internal resistor into node j for $j=k$ or away from any node j for $j \neq k$. Applying KCL at each node implies that $\Delta i_j \geq 0$ if $j=k$ and $\Delta i_j \leq 0$ if $j \neq k$. Since this is true at any operating point, $f(v)$ is monotonic as claimed[31]. To investigate Lipschitz continuity, pick any two inputs v and $\hat{v}$ and use the max norm on $\mathbb{R}^n$. Let p be the number of resistors in $\mathcal{N}_R$. Let K_k be the Lipschitz

---

[29] Both a monotonically increasing function and a monotonically decreasing function, as defined here, include constant functions. Strict monotonicity is used to imply strict inequalities.

[30] Uniqueness requires strict incremental passivity as supplied by definition 4-1. Existence requires that for each resistor, $i \rightarrow \infty$ as $v \rightarrow \infty$, which is implied by the definition.

[31] If the function $f(v)$ is assumed to be C^1, monotonicity can be shown by observing that $[J(f)]_{(v)}$ is an M-matrix [43]. Even more structure of $\mathcal{N}_R$ becomes apparent through the known properties of M-matrices [44], and this structure has been useful in bounding the behavior of RC circuits [30] [33] [34] [45].

constant for the constitutive relation $i_k = g_k(v_k)$ of each internal resistor R_k for $n \le k \le n+p-1$. Let $K = p(\max\{K_k, n \le k \le n+p-1\})$. By the incremental voltage minimax theorem, $|\Delta v_k| \le \|v - \hat{v}\|$ for any internal branch voltage v_k, $n \le k \le n+p-1$. Therefore, $\|f^R(v) - f^R(\hat{v})\| = \max\{\hat{i}_j - i_j, 0 \le j \le n-1\} \le p(\max\{|\Delta i_k|, n \le k \le n+p-1\}$ by applying KCL at each terminal node and observing none can be connected to more than p resistors. $|\Delta i_k| \le K_k |\Delta v_l|$ $\forall$ $n \le k \le n+p-1\}$ so $\|f^R(v) - f^R(\hat{v})\| \le p(\max\{K_k, n \le k \le n+p-1\})(\max\{|\Delta v_k|, n \le k \le n+p-1\}) \le K\|v - \hat{v}\|$. ∎

If bounds are given for all of the terminal voltages, bounds on the terminal currents can be calculated by simply evaluating the constraint function f^R for subnetwork $\mathcal{N}_R$ using different combinations of voltage endpoints. For example, an upper bound on any terminal current i_k is calculated by using the upper voltage bound for node k and the lower voltage bound for all other terminal nodes. In practice, the evaluation of a component of the function f^R usually involves the addition of the current contributions from a few resistors with known terminal voltages.

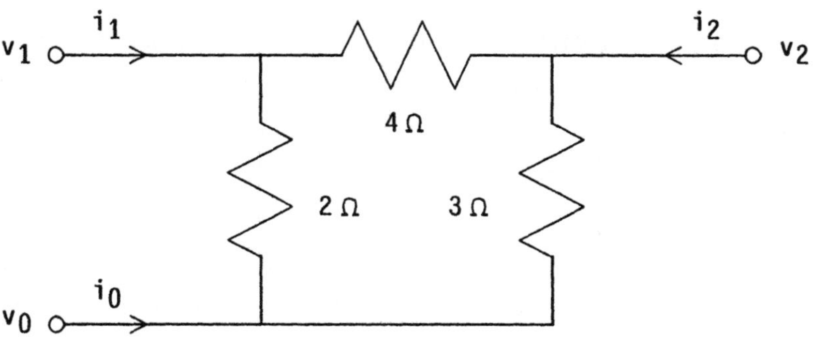

Figure 4-8: The terminal constraints of $\mathcal{N}_R$ are monotonic.

The simple circuit in figure 4-8 serves to illustrate the use of a monotonic terminal constraint function in bounding the terminal currents, given bounds on terminal voltages. In the example, all terminal voltages are known to fall in the interval [0,1] volts. To find the maximum value for the current i_1, the maximum value for v_1 is used along with the minimum values for v_0 and v_2. Given these extreme voltages, the current through the 2Ω resistor is 0.5 amp and that through the 4Ω resistor is 0.25 amp, both contributing positively to i_1. As a result, the maximum value for the current i_1 is 0.75 amp. Note that the absence of internal nodes in $\mathcal{N}_R$ makes the calculation of $\mathbf{f}^R(\mathbf{v})$ trivial.

4.2.2 Internal Elements

The terminal constraint $\mathbf{i}=\mathbf{f}^R(\mathbf{v})$ of $\mathcal{N}_R$ is a monotonic (suitably defined) function, but the effect of internal elements on the function must also be identified. For simple bounding algorithms, substitution of any internal resistor with a "larger" resistance must be guaranteed to have a monotonic effect on the function $\mathbf{f}^R$, i.e., it must always reduce some terminal currents and increase others given that the terminal voltages are fixed.

In general, no simple monotonic relationship between internal elements and terminal constraints exists. Consider the linear network pictured in figure 4-9. For a fixed set of terminal voltages and a fixed conductance G_1, i_1 is a monotonically decreasing function of the node voltage v at the internal node. Since $v=(v_1G_1+v_2G_2)/(G_0+G_1+G_2)$, the derivative of v with respect to G_2 has the same sign as $v_2(G_0+G_1)-v_1G_1$. In this circuit the effect of G_2 on the terminal current i_1 is always monotonic, but the direction depends on the values of the other resistors[32]. Unfortunately, for circuits that include nonlinear resistors, the currents can even be nonmonotonic functions of the resistors. An example of such a circuit with only two terminals is presented in the next subsection.

A reasonable solution to the element monotonicity problem is to add the constraint that all internal resistors must be grouped into one-ports whose terminals are two of the external nodes. Figure 4-10 pictures a network $\mathcal{N}_R$,

---

[32] If the property of monotonicity with unknown polarity were true in general, a bound could be obtained by comparing the currents produced by all possible combinations of extreme resistor values. Such a search method is considered in detail for the capacitor subnetwork $\mathcal{N}_C$.

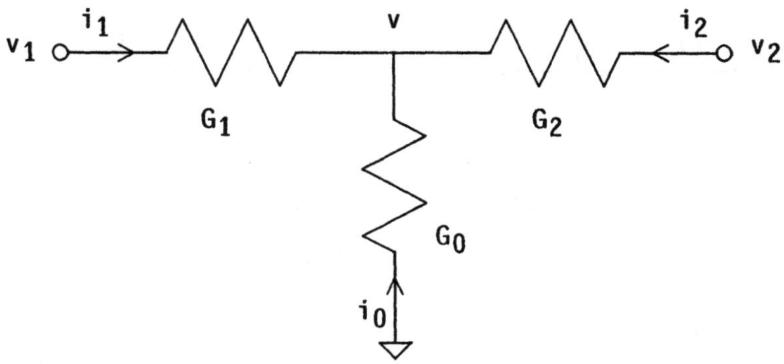

Figure 4-9: Example of $\mathcal{N}_R$ with complex dependence on resistor G_2.

containing four such one-port groups, that satisfies the constraint. Each of these one-port resistors is called a "one-port group," and is assigned a reference direction. This restriction allows the dependence of the terminal constraints on internal elements to be partitioned into two effects. The first effect is that of the element on its one-port group. In the next subsection, conditions that guarantee each of the one-port groups is a monotonic function of its internal resistors are considered. The second effect is that of a one-port group on the terminal constraints of $\mathcal{N}_R$. Each terminal current is a current-wise monotonic[33] function of each one-port group, with the sign of the relationship depending only on the reference direction of the one-port group.

To explore the effect of a one-port group on the entire subnetwork, consider the one-port group labeled R_1 in figure 4-10. First note that a one-port group is a passive and incrementally passive resistor because the passivity and incremental passivity properties of *R*-elements are closed under composition. As a result, the partial orderings defined for resistors in the last section can be used for one-port groups as well. Assume that all terminal voltages are fixed and that only the resistance of R_1 changes. Since all terminal voltages are fixed,

---

[33]Current-wise monotonicity may seem like a strange notion, but it is merely a notational convenience. If the simple partial ordering of resistors were used, the sign of the monotonic relationships would depend on terminal voltages.

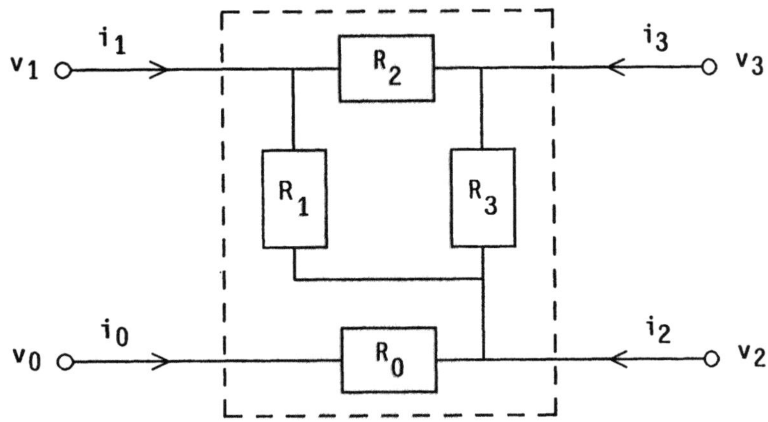

Figure 4-10: All internal nodes of $\mathcal{N}_R$ are inside one-port groups.

the currents through all of the other one-port groups remain the same. Only the current through R_1 changes, implying that only the terminal currents i_1 and i_2 change. Since its voltage drop is fixed, increasing R_1 in terms of the current-wise ordering is guaranteed to increase its current. In addition, the terminal currents i_1 and i_2 have a monotonic dependence on the current passing through R_1 as a result of the KCL constraints at each terminal. Therefore, if R_1 is increased in the current-wise ordering, i_1 is increased if the reference direction of R_1 points away from terminal one and decreased if it points towards the terminal. The same is true for i_2. In general, a terminal current is a current-wise monotonically increasing function of each one-port group that is connected to its terminal and has a reference direction pointing away from its terminal.

The one-port grouping constraint is not as severe as it might first seem because it only rules out complex multi-port resistor networks with negligible internal capacitance. The structure of most integrated circuits makes the one-port groupings likely, and where they do not exist, small capacitors can be added[34]. If a grounded capacitor is added to every node, the subnetwork $\mathcal{N}_R$

[34] A small capacitor allows the solution of what is essentially a d.c. problem to be obtained with the more powerful relaxation techniques considered later for transient analysis.

becomes a network as pictured in figure 4-9, where all the one-port groups are either a single resistor or a number of individual resistors in parallel. The circuit in figure 4-6 provides an example that requires the addition of a small grounded capacitor for the resistor subnetwork $\mathcal{N}_R$ to meet the one-port grouping constraint.

4.2.3 One-Port Groups

Now the one-port groups of figure 4-10 are considered. Constraints on the groups are needed to insure that they have a resistance that is a monotonically increasing function of the resistance of each of their internal resistors. For any one-port network made by interconnecting linear resistors, the total resistance is a monotonically increasing function of the value of each internal resistor [46]. Surprisingly, this monotonicity property cannot be generalized to one-port networks constructed with nonlinear resistors, even those that are both passive and strictly incrementally passive, unless further constraints are placed on the network. However, monotonicity is guaranteed for any resistor that is included with all nonlinear resistors in a series-parallel subgroup. Such a nonlinear grouping restraint is therefore required for any resistor that is incompletely specified, i.e., bounded, due to device uncertainty or model simplifications.

One-port groups made of monotone resistors exhibit nonmonotone sensitivity whenever their incremental and large scale behaviors, with respect to branch variables, differ in sign. A more complete investigation of this behavior is presented in [47]. An example from [47] is provided here to illustrate nonmonotonic behavior. The circuit in figure 4-11 contains ideal diodes that turn on abruptly[35] at 0.6 volts, and linear resistors. Assume R_0 is a linear resistor of value $r_0 = 1\Omega$ and the port voltage $v_p = 1.0$ volts. Since the diodes are turned on at this operating point, $v_a = 0.4$, $v_b = 0.6$, and the port current is therefore $0.8 - (0.2/r_0)$. If r_0 is increased only slightly, the equation remains valid and the port current actually rises.

There is a large, simple class of nonlinear one-port groups that do exhibit monotone sensitivity with respect to the resistance of some internal branch called b_0. If branch b_0 is included along with all nonlinear resistors in a series-parallel subgroup, the entire group has monotone sensitivity with respect to the

---

[35] With a little modification, the diode i-v curves can be made to satisfy the R-element definition, and the operation of the circuit will be essentially unchanged.

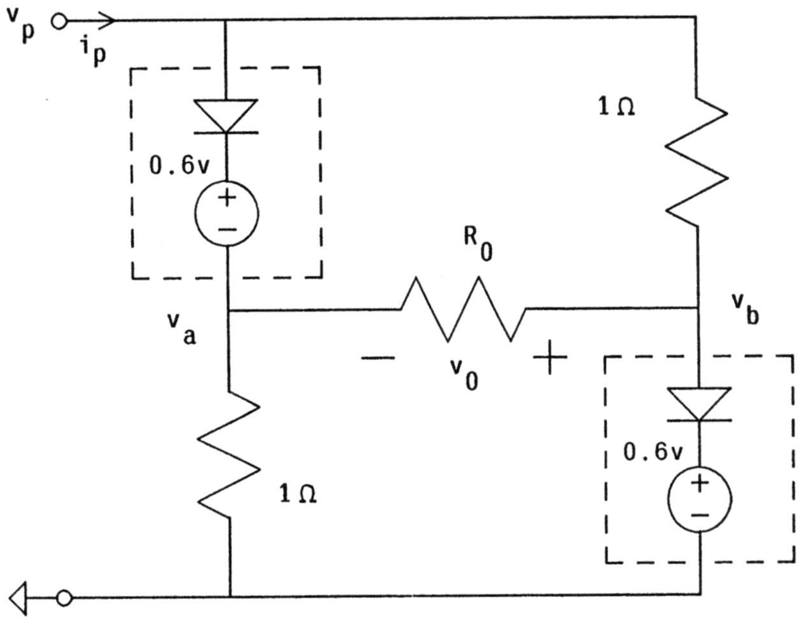

Figure 4-11: A simple bridge circuit that is nonmonotonic with respect to R_0.

resistance of b_0. All incompletely specified resistors in a one-port group are therefore required to be branches that satisfy this constraint. A detailed derivation of this constraint is presented in this subsection.

Consider a one-port group $\mathcal{R}$ constructed from resistors (R-elements) R_1 to R_m and an as yet undefined branch b_0. One-port groups R_p and $\hat{R}_p$ are formed by replacing branch b_0 in $\mathcal{R}$ with resistors R_0 and $\hat{R}_0$, respectively. An arbitrary voltage source is placed across the ports of both R_p and $\hat{R}_p$. Associated reference directions are then assigned to all branches in R_p so that v_j and i_j for $j=0$ to m are nonnegative. The variables v_p and i_p are also nonnegative and are assigned as pictured in figure 4-12. Reference directions for branches of $\hat{R}_p$ are assigned to agree with those of corresponding branches of R_p, and the branch variables of $\hat{R}_p$ are labelled $\hat{v}_j$ and $\hat{i}_j$. The differences $\hat{v}_j - v_j$ and $\hat{i}_j - i_j$ are denoted Δv_j and Δi_j respectively.

Definition 4-7: We say that the one-port group $\mathcal{R}$ has monotone sensitivity with respect to branch b_0 iff $\hat{i}_p \leq i_p$ ($\Delta i_p \leq 0$) for all port voltages v_p and for all such one-port groups $\hat{R}_p$ and R_p in which $\hat{R}_0 \geq R_0$. Equivalently, $\hat{R}_p \geq R_p$ whenever $\hat{R}_0 \geq R_0$.

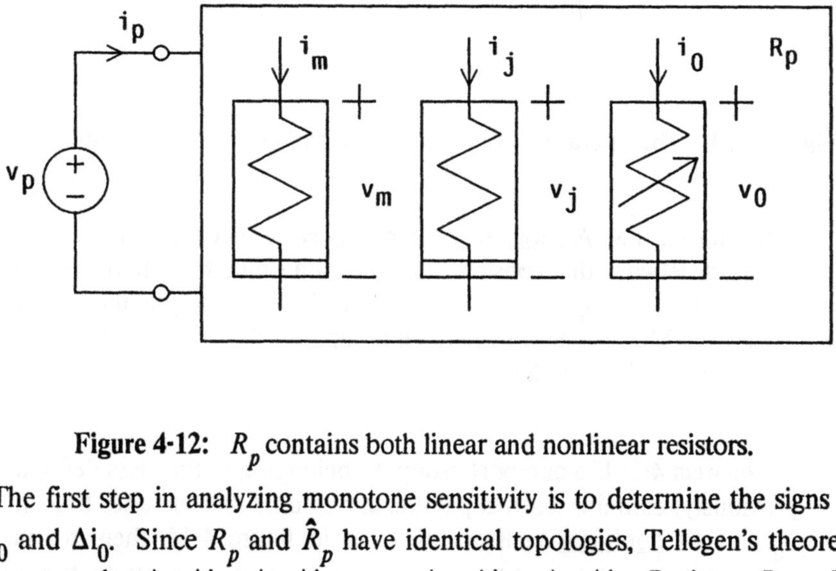

Figure 4-12: R_p contains both linear and nonlinear resistors.

The first step in analyzing monotone sensitivity is to determine the signs of Δv_0 and Δi_0. Since R_p and $\hat{R}_p$ have identical topologies, Tellegen's theorem guarantees that $\Delta v_0 \Delta i_0 + \Delta v_1 \Delta i_1 + \ldots + \Delta v_m \Delta i_m = \Delta v_p \Delta i_p$. Resistors R_1 to R_m are identical in both networks and are incrementally passive, so $\Delta v_j \Delta i_j \geq 0$, $j = 1 \ldots m$. (Of course, Δv_j and Δi_j could both be zero). Also, the same port voltage v_p is placed across both one-port groups, so $\Delta v_p = 0$. As a result, $\Delta v_0 \Delta i_0 \leq 0$. Hence, the point $(\hat{v}_0, \hat{i}_0)$ must lie in the shaded region in figure 4-13. Since the point $(\hat{v}_0, \hat{i}_0)$ must also satisfy the element constraint $\hat{v}_0 = \hat{f}_0(\hat{i}_0)$, and $\hat{R}_0 \geq R_0$, it follows that $\Delta v_0 \geq 0$ and $\Delta i_0 \leq 0$.

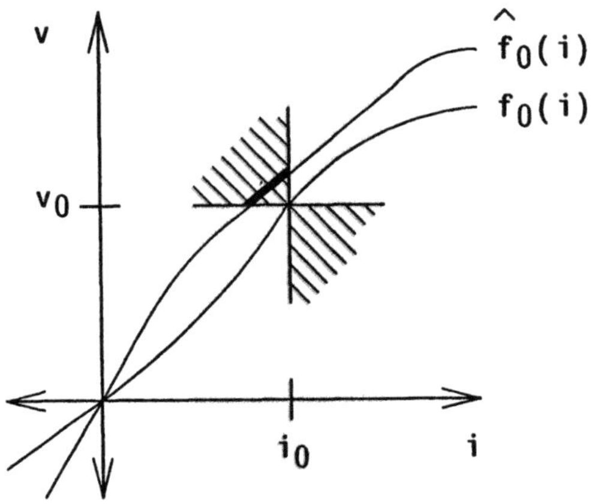

Figure 4-13: The operating point lies on the darkened portion of the curve.

Definition 4-8: A series-parallel group is defined recursively as a one-port network that consists of either: a) a single branch, or b) two series-parallel groups in series or in parallel. For example, the circuit in figure 4-14 is not a series-parallel group, but the nonlinear resistors in it form a series-parallel group.

Theorem 4-2: If a one-port group $\mathcal{N}$ (belonging to the class defined previously) contains a series-parallel subgroup $\mathcal{S}$ that includes branch b_0 and all nonlinear resistors in $\mathcal{N}$, as in figure 4-14, then $\mathcal{N}$ has monotone sensitivity with respect to b_0. ($\mathcal{S}$ may include linear resistors as well.)

For the proof of the preceding theorem it is useful to define the quantity $x_j = i_j \Delta v_j - v_j \Delta i_j$ for each branch in $\mathcal{N}$. Note that, with the sign conventions adopted here, Tellegen's theorem guarantees $x_p = x_0 + x_1 + \ldots + x_m$. We test $\mathcal{N}$ for monotonicity by driving it with a voltage source, so $x_p = -v_p \Delta i_p$ because

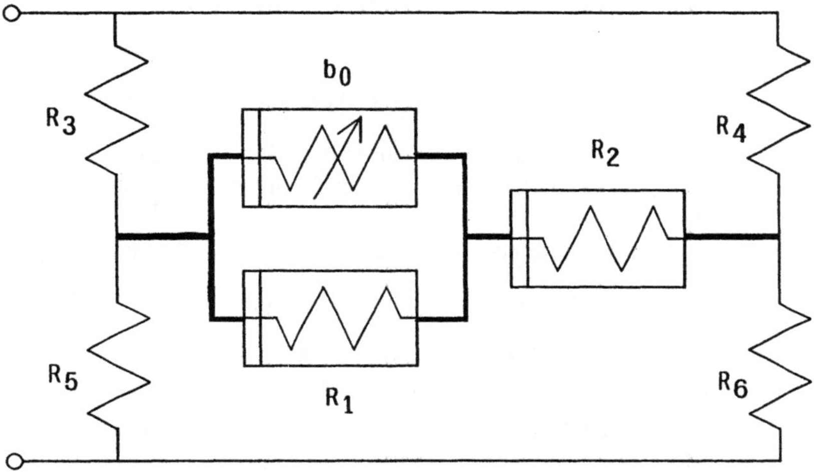

Figure 4-14: Branches b_0, R_1, and R_2 form a series-parallel subgroup.

$\Delta v_p = 0$. Thus, x_p has the opposite sign from Δi_p. Therefore, since $\hat{R}_0 \geq R_0$ by assumption, all that must be shown to prove monotonicity is that $x_0 + x_1 + \ldots + x_m \geq 0$. From the general results presented so far, it only follows that $x_0 \geq 0$ since $\Delta v_0 \geq 0$ and $\Delta i_0 \leq 0$.

Proof: If R_j, $1 \leq j \leq m$, is a linear resistor, then $v_j = r_j i_j$. Since $\Delta v_j = r_j \Delta i_j$, $x_j = i_j \Delta v_j - v_j \Delta i_j = i_j \Delta i_j r_j - i_j \Delta i_j r_j = 0$. Thus, the contribution of all linear resistors in the sum $x_0 + x_1 + \ldots x_m$ can be ignored. Next, assume that $\mathcal{S}$ includes b_0 and at least one other resistor. A series-parallel subnetwork of $\mathcal{S}$, R_{k_1}, is either in series or parallel with b_0 and hence R_{k_1} and b_0 form a resistive one-port R_{p_1}. The one-port resistor R_{k_1} is a series-parallel network containing R-elements, and is therefore an R-element itself. Since $v = f_{p_1}(i) = f_{k_1}(i) + f_0(i)$ in the series case and $i = g_{p_1}(v) = g_{k_1}(v) + g_0(v)$

in the parallel case, it follows that $\hat{R}_{p_1} \geq R_{p_1}$. Using the previous results, $\hat{R}_{p_1} \geq R_{p_1}$ guarantees that $\Delta v_{p_1} \geq 0$ and $\Delta i_{p_1} \leq 0$, i.e., $x_{p_1} \geq 0$. Also, from Tellegen's theorem, x_{p_1} is the sum of x_0 and the x_i associated with each resistor in R_{k_1}. If there are still more resistors in $\mathcal{I}$, a series-parallel subnetwork of $\mathcal{I}$ must be either in series or parallel with R_{p_1}. Call this subnetwork R_{k_2}, and consider R_{k_2} and R_{p_1} as forming a resistive one-port R_{p_2}. As before, $\hat{R}_{p_2} \geq R_{p_2}$ and $x_{p_2} \geq 0$, i.e., the sum of the x_i associated with each resistor in R_{p_2} is nonnegative. Proceed in this manner until all resistors in $\mathcal{I}$ are exhausted.. By induction, the component of x_p that comes from b_0 and all resistors in $\mathcal{I}$ cannot be negative. Since all other components of x_p are zero, $x_p \geq 0$ and $\mathcal{R}$ must have monotone sensitivity as claimed. ∎

In a MOS circuit model of intermediate accuracy, each of the one-port groups inside the resistor subnetwork $\mathcal{N}_R$ can be a fairly complex combination of resistors, provided that all resistors specified by resistance intervals and all nonlinear resistors can be placed in a single series-parallel subgroup. A common special case of this restriction is when the one-port group is itself a series-parallel group. With this restriction on the topology of the one-port group, it is guaranteed to be a monotonic function of all its internal resistors. As a result, intervals for each internal resistor can be readily transformed into intervals on the one-port group through composition of two sets of resistances. An upper bound on a one-port group resistor is calculated using the maximum resistance of each internal element. Likewise, a lower bound on a one-port group resistor is calculated using the minimum resistance of each internal element.

4.2.4 Resistor Subnetwork Summary

The results of this section are summarized in a definition and a theorem. The definition states the restrictions placed on the topology of the resistor subnetwork $\mathcal{N}_R$ and the theorem states all of its important properties that are used later.

Definition 4-9: The resistor subnetwork $\mathcal{N}_R$ is a multi-terminal resistor constructed from R-elements. Each internal element is part of a one-port group whose two terminals are external terminals of $\mathcal{N}_R$. In each of these one-port groups, all of the internal elements that are incompletely specified, i.e., characterized with bounds, that are nonlinear, or that are both, are contained, possibly with other elements, in a single series-parallel subgroup.

Theorem 4-3: The resistor subnetwork $\mathcal{N}_R$ defines a unique mapping $i = f^R(v)$ satisfying a Lipschitz condition. Any component of i, i_j for $0 \leq j \leq n-1$, is a monotonically increasing function of v_j, a monotonically decreasing function of v_k for $k \neq j$, $0 \leq k \leq n-1$, a current-wise monotonically increasing function of any one-port group connected to terminal j with a reference direction pointing away from the terminal, a current-wise monotonically decreasing function of any one-port group connected to terminal j with a reference direction pointing towards the terminal, and a constant function of any other one-port group. Furthermore, every one-port group is a monotonically increasing function of all its internal incompletely specified resistors.

Proof: The first statement of the theorem and monotonicity with respect to terminal voltages are special cases of theorem 4-1. Monotonicity of the one-port groups with respect to internal incompletely specified resistors is a direct result of theorem 4-2 along with definition 4-9. Current-wise monotonicity of i_j with respect to the one-ports connected to terminal j follows from an application of KCL at the terminal and the definition of a current-wise ordering given as definition 4-2. Since i_j is completely determined by these currents, which are only determined by the connected one-ports and their terminal voltages, i_j is not affected by other one-ports groups. ∎

In simple MOS circuit models, time-varying resistors are often used to model transistors. Even though the definition for resistors used here does not allow them to be time-varying, these models can be accommodated by viewing a simulation with such a model as a sequence of simulations over short time periods, each with different circuit models. Bounds on the behavior of each can be pieced together through bounds on the final and initial states. In such a simulation, the resistance of each one-port group must be recalculated a number of times, so its complexity becomes more important. In this case efficiency is improved if the one-port groups are entirely series-parallel, which is often the case in practice when resistors are used as a simple model for transistors.

4.3 The Transistor Subnetwork

This section considers the subnetworks of the MOS circuit model $\mathcal{N}_N$ and $\mathcal{N}_P$ containing all of the n-channel and p-channel transistor elements respectively. $\mathcal{N}_N$ is considered in detail first, and then the differences between the two are discussed. Just as when considering the resistor subnetwork, reasonable restrictions on the subnetwork that facilitate bounding are derived. It is shown that the desirable constraints are similar to those on the resistor subnetwork, if a transistor model is treated as if its channel were simply a resistor. All transistors must be contained in series-parallel "one-port" groups. In addition, the gate terminals of all the transistors except those in a "diode" configuration must be connected to external terminals of the subnetwork. Substrate terminals are treated like gate terminals and, for simplicity, it is assumed that all are tied to a single external node, as is almost always the case in practice.

4.3.1 Comparison with the Resistor Subnetwork

The transistor subnetwork $\mathcal{N}_N$, as pictured in figure 4-15, affects the rest of the circuit through its terminal constraint $\mathbf{i}=\mathbf{f}^N(\mathbf{v})$ just as the resistor subnetwork does. Again $\mathbf{i}$ and $\mathbf{v}$ represent vectors whose components are the terminal currents and voltages respectively. The reasons for considering the terminal currents to be the dependent variables are similar to those given for the resistor subnetwork. An additional reason applying only to MOS transistor subnetworks is that MOS transistors cannot be expressed completely in a current-controlled manner, but they can be described using voltages as the independent variables. The existence of a unique solution (and thus a well defined function f^{N}) under the constraints imposed in this section is discussed later.

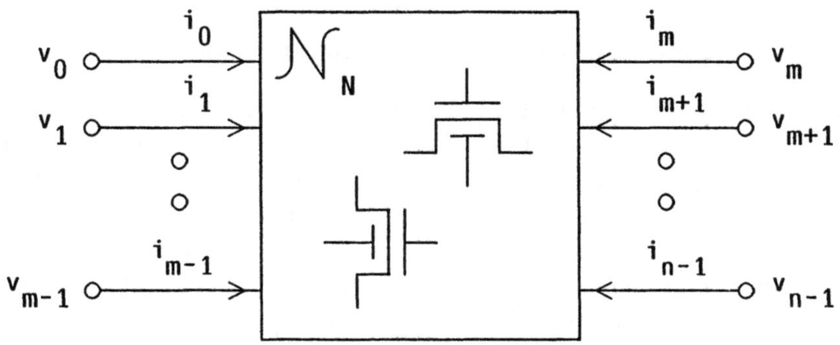

Figure 4-15: The n-channel transistor subnetwork $\mathcal{N}_N$.

For bounding algorithms, simple monotonic relations between terminal variables and device characteristics are needed. Since a nonlinear one-port resistor is a special case of a transistor with nonfunctioning gate and substrate terminals according to the definitions in section 4.1, the transistor subnetwork is at least as difficult to deal with. It is reasonable to start with the constraints imposed on the resistor subnetwork as they should be a subset of the constraints on a more general subnetwork. Therefore, it is assumed that all transistors in $\mathcal{N}_N$ are grouped into "one-ports," where the notion of being a one-port is defined by ignoring the gate and substrate transistor terminals. The two terminals of each one-port must be connected to terminals of $\mathcal{N}_N$. As with $\mathcal{N}_R$, these one-ports are called "one-port groups," and each is assigned a reference direction. Furthermore, since the resistance of all transistors is nonlinear, the "one-ports" must be series-parallel groups[36], where series-parallel is defined for N-elements as discussed in the next subsection. Roughly speaking, series-parallel groups of N-elements are the same as series-parallel groups of resistors if their gate and substrate terminals are ignored. The reference direction of each

---

[36] A linear resistor is a special case of the transistor as defined here, but generalizing constraints to specially treat such a case is of little practical value.

transistor in a one-port group is assigned so that it is consistent with that of its group. The reference direction of a transistor is the same as the reference direction of its drain current, and it is assumed that the choice of drain and source are fixed, i.e., the choice does not change as a function of circuit voltages.

4.3.2 Series and Parallel Transistor Compositions

To investigate the properties of series-parallel groups of transistors, a generalized definition for an *N*-element is introduced. The definition is similar[37] to that of the original except that the restriction of having a single gate terminal is removed. A single substrate terminal is assumed just for simplicity, as series-parallel transistor groups normally have a common substrate connection. All of the monotonic properties are the same except that they are extended to all gate terminals. Partial orderings for general *N*-elements are also defined as before.

> **Definition 4-10**: A <u>general *N*-element</u> is a (g+3)-terminal resistive element, as pictured in figure 4-16, where $i_{G1} = \ldots = i_{Gg} = i_B = 0$ and $i_D = -i_S = F(v_{G1}, \ldots v_{Gg}, v_D, v_S, v_B)$, $(g \geq 0)$. The function $F: \mathbb{R}^{g+3} \rightarrow \mathbb{R}$ evaluates to zero if $v_D = v_S$. F is a strictly monotonically increasing function of v_D and a strictly monotonically decreasing function of v_S. $|F|$ is a monotonically increasing function of all v_{Gj} for $j = 1$ to g and a monotonically decreasing function of v_B. Given two (g+3)-terminal *N*-elements N_1 and N_2, we say that " $N_1 \geq N_2$ " iff $[F_1(v_{G1}, \ldots v_{Gg}, v_D, v_S, v_B) - F_2(v_{G1}, \ldots v_{Gg}, v_D, v_S, v_B)](v_D - v_S) \geq 0$ $\forall v_{G1}, \ldots v_{Gg}, v_D, v_S$, and v_B. We say that "$N_1 \geq N_2$ current-wise" iff $F_1(v_{G1}, \ldots v_{Gg}, v_D, v_S, v_B) \geq F_2(v_{G1}, \ldots v_{Gg}, v_D, v_S, v_B)$ $\forall v_{G1}, \ldots v_{Gg}, v_D, v_S$, and v_B.

A general *N*-element with g=1 describes a model for a single n-channel transistor, while a general *N*-element with g=0 describes a model for a

---

[37] Lipschitz continuity of a series-parallel group does not follow from Lipschitz continuity of each internal element without further reasonable restrictions on the transistor model. Rather than add such restrictions, Lipschitz continuity is simply required for each series-parallel group.

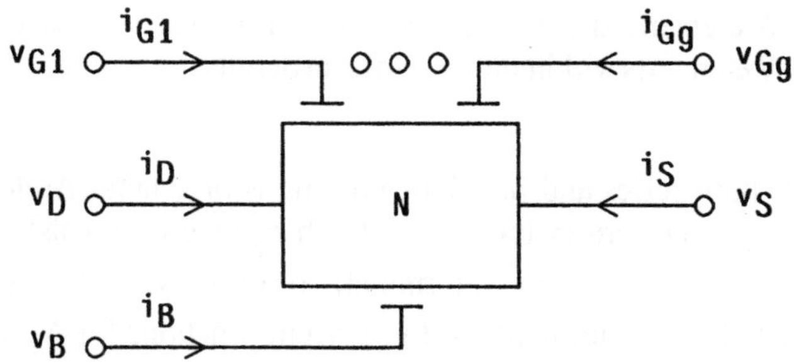

Figure 4-16: General *N*-elements model NMOS pullups and transistor groups.

depletion load pullup[38]. The following theorem shows (along with an inductive argument) that a general *N*-element with g>1 can describe a model for a group of n-channel transistors provided they are connected in series and parallel configurations. Such networks, possibly modeling the pulldown networks of MOS logic gates, possess the same basic monotonic terminal properties as individual transistors. Furthermore, the theorem shows that a series-parallel group is a monotonic function of its individual transistors with respect to either partial ordering. For example, if the width of any transistor in a series-parallel group is increased, the magnitude of the current through the group will also increase for a given set of terminal voltages.

Definition 4-11: Two general *N*-elements are said to be in series if they are connected as in the left circuit in figure 4-17, with a common substrate terminal, and only the two pictured connections to terminal x. Two general *N*-elements are said to be in parallel if they are connected as in the right circuit in figure 4-17, with a common substrate terminal.

---

[38] Any transistor connected in a "diode" configuration satisfies the definition of a general *N*-element as a consequence of the constraints in definition 4-3.

Theorem 4-4: Two general N-elements connected in either series or parallel, form another general N-element. Furthermore, the resulting general N-element is a monotonic function of each of its internal general N-elements within either partial ordering.

Proof: Both series and parallel connections of general N-elements N_1 (with g_1 gate terminals) and N_2 (with g_2 gate terminals) form a $(3+g_1+g_2)$-terminal ((g+3)-terminal) resistor N. To simplify arguments throughout, treat the two current functions for N_1 and N_2 as if they had all g gate voltages as inputs, each with no dependence on the ones from the other element. Consider first the parallel case. By the definitions of N_1 and N_2, $i_{G1}= .. =i_{Gg}=i_B=0$. Using KCL and the definition for N_1 and N_2, it follows that $i_D=i_{D1}+i_{D2}=-i_{S1}-i_{S2}=-i_S$. If $v_D=v_S$, $i_D=i_{D1}+i_{D2}=0$. Furthermore, i_{D1} and i_{D2} always have the same sign because N_1 and N_2 have the same voltage polarity. As a result, $|i_D|=|i_{D1}|+|i_{D2}|$ as well. The fact that the current function $i_D=F(v_{G1}, .. v_{Gg}, v_D, v_S, v_B)$ for N is simply the sum of the current functions of N_1 and N_2 implies that the function exists, is unique, is strictly monotonically increasing in v_D, and is strictly monotonically decreasing in v_S. It also implies that N is current-wise monotonic in N_1 and N_2. The fact that the current magnitudes also add implies monotonicity in the gate and substrate voltages, as well as simple monotonicity of N in N_1 and N_2. Consider second the series case. By the definitions of N_1 and N_2, $i_{G1}= .. =i_{Gg}=i_B=0$. Using KCL and the definition for N_1 and N_2, it follows that $i_D=i_{D1}=-i_{S1}=i_{D2}=-i_{S2}=-i_S$. To investigate existence and uniqueness of the current function, pick a set of input voltages $v_{G1}, .. v_{Gg}, v_D, v_S, v_B$ such that, without loss of generality, $v_D \geq v_S$. Let $i_X=i_{D1}-i_{D2}$, which is a continuous and strictly monotonically decreasing function of v_X if it is assumed for now that v_X is controlled by a grounded voltage source. $i_X>0$ if $v_X=v_S$ and $i_X<0$ if $v_X=v_D$, so there exists a unique v_X for which $i_X=0$. The unique current function of either N_1 or N_2 then implies existence and uniqueness of the current function for N. The fact that the current function evaluates to zero if $v_D = v_S$ can then be verified by direct substitution, setting $v_X = v_D$, as a result of uniqueness. Now consider the monotonic properties of N. Choose any operating point $\{v_{G1}, \ldots v_{Gg}, v_D, v_S, v_B\}$, along with its unique internal voltage v_X and current i_D. First compare this to another operating point $\{v_{G1}, \ldots v_{Gg}, \hat{v}_D, v_S, v_B\}$, along with its unique internal voltage $\hat{v}_X$ and current $\hat{i}_D$, such that $\hat{v}_D > v_D$. Assume $\hat{i}_D \leq i_D$. Because N_2 has a current that is strictly monotonic in v_X, this implies that $\hat{v}_X \leq v_X$. Due to the monotonicity of N_1, though, $\hat{v}_X > v_X$ if $\hat{i}_D \leq i_D$ so this leads to a contradiction. As a result, $\hat{i}_D > i_D$ so i_D is a strictly monotonically increasing function of v_D. By symmetry, $i_S = -i_D$ is a strictly monotonically increasing function of v_S so i_D is a strictly monotonically decreasing function of v_S. Now compare the first operating point to another one $\{\hat{v}_{G1}, \ldots v_{Gg}, v_D, v_S, v_B\}$, along with its unique internal voltage $\hat{v}_X$ and current $\hat{i}_D$, such that $\hat{v}_{G1} \geq v_{G1}$. Without loss of generality assume v_{G1} is applied to element N_1. If $v_D \geq v_S$, assume that $\hat{i}_D < i_D$. The element currents must be nonnegative so the definition of N_2 implies that $\hat{v}_X < v_X$ and the definition of N_1 implies that $\hat{v}_X > v_X$. If $v_D < v_S$ and it is assumed that $\hat{i}_D > i_D$, a similar contradiction exists. As a result, $|\hat{i}_D| \geq |i_D|$ and the current magnitude is a monotonically increasing function of the gate voltages. The substrate terminal is the same as a gate terminal except that the polarity is opposite. The fact that it is connected to both elements only reinforces its effect. The argument for monotonicity of N with respect to N_1 and N_2 is similar to monotonicity in gate voltage. Pick any operating point $\{v_{G1}, \ldots v_{Gg}, v_D, v_S, v_B\}$, and consider, without loss of generality, increasing N_1. If $v_D > v_S$, i_D cannot fall, as doing so would lead to a contradiction in the movement of v_X. If $v_D < v_S$, the effect on i_D depends on the partial ordering used for N_1. If a current-wise ordering is used, i_D still cannot fall for identical reasons. If a simple ordering is used, i_D cannot rise as the polarity of the effect is reversed. For either partial ordering, monotonicity of N then follows. ∎

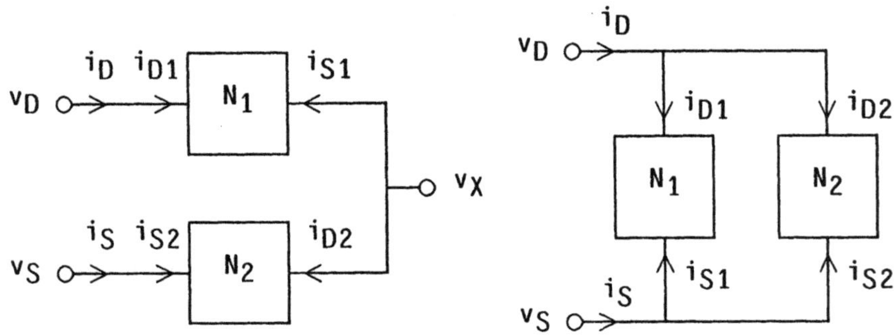

Figure 4-17: Series and parallel connections of general N-elements.

Since N-elements are special cases of general N-elements, definition 4-11 can be used to precisely define a series-parallel group of N-elements. Theorem 4-4 then implies that a series-parallel group is a general N-element. Requiring that each one-port group is series-parallel implies that it is also a general N-element. As a result, it makes sense to refer to the drain and source of a one-port group. The definition of series and parallel used here forces all the gate terminals of transistors, except those used to create "diode" configurations, to be connected to external terminals of $\mathcal{N}_N$, as they cannot be connected to any internal nodes. This simplification is necessary to generate monotonic terminal constraints, and it is very mild in practice since gate terminals have significant capacitance. Recall that any node connected to a capacitor is an external terminal of $\mathcal{N}_N$. Small grounded capacitors can always be added to meet this constraint.

4.3.3 Terminal Constraints and Internal Elements

Assume for now that the gate terminals of all transistors, as well as the substrate terminals, are fixed in voltage and are not treated as terminals. $\mathcal{N}_N$ is simply a number of general N-elements whose drains and sources are connected to external terminals of $\mathcal{N}_N$ and whose gates and substrates are fixed in voltage.

By applying KCL at each terminal of $\mathcal{N}_N$, as in the case of $\mathcal{N}_R$, one finds not only that there is a unique mapping $\mathbf{i}=\mathbf{f}^N(\mathbf{v})$, but also the same monotonic terminal constraints found in $\mathcal{N}_R$. Each terminal current is a monotonically increasing function of its corresponding terminal voltage and a monotonically decreasing function of all other terminal voltages.

Of course the gate terminals are not at fixed voltages, but instead are connected to external terminals of $\mathcal{N}_N$. Existence and uniqueness of the terminal currents still applies, but the terminal constraints are not monotonic. Consider the example of figure 4-18. Overestimating the voltage v_1 can either cause the current i_2 to be overestimated or underestimated, depending on the relative transistor characteristics.

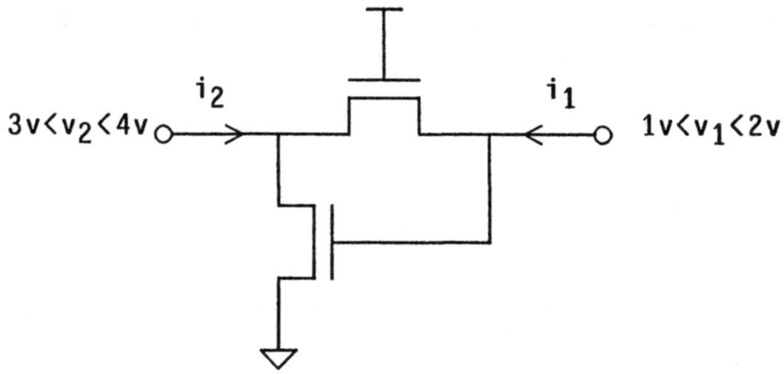

Figure 4-18: The terminal constraints of $\mathcal{N}_N$ are not necessarily monotonic.

The nonmonotonicity of the terminal constraint created by the gates of MOS transistors poses a fundamental problem that arises in bounding analysis. Although the situation pictured in figure 4-18 might be unlikely, ruling it out would impose fundamental restrictions on the type of circuit that could be considered, not just small variations in the type of model used. The circuit already has no internal nodes, so the addition of arbitrarily small capacitors will not further simplify the transistor subnetwork topology. An example of a real circuit that poses such a problem is the Schmitt trigger discussed in section 3.3.

Fortunately there is a trick that can be used to transform such a circuit into an equivalent one that does exhibit monotonicity. The cost is that the correlation between multiple uses of the same terminal voltage on transistor gates is ignored. A bound is generated, not from the solution of a particular operating point of the circuit, but rather from the operating point of an equivalent one. The new circuit is made by creating an additional external terminal for each transistor gate, as pictured for the circuit of the previous example in figure 4-19. Now various terminal voltages are duplicated, and the upper bound might be used in one instance while a lower bound is used in others. In figure 4-19 an upper bound on i_2 is calculated by evaluating the circuit with v_{1A} at its minimum and v_{1B} at its maximum.

One more topology constraint must be added to remove internal feedback paths. A terminal that is connected to the source or drain of a one-port group cannot also be connected to one of the gate terminals associated with that group, except for "diode" connections[39]. The need for this constraint is discussed further in section 4.5. Roughly speaking, when calculating a single terminal current of $\mathcal{N}_N$, it would be better not to have to use two instances of the corresponding terminal voltage.

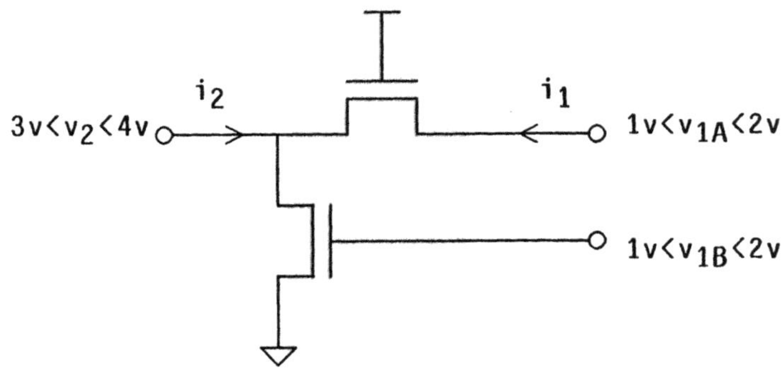

Figure 4-19: A modified circuit can be used to calculate bounds.

Another new situation that arises for the terminal constraints of $\mathcal{N}_N$ is that the direction of monotonicity in some terminal voltages is a function of other

---

[39]Note that as long as diode configurations are allowed, this constraint only affects model complexity. The exception is needed, though, as one-port groups cannot be reduced beyond a single transistor.

voltages. Since only the polarity of voltages across transistors[40] determines the direction of monotonicity, there is another trick that can further simplify[41] treatment of $\mathcal{N}_N$. The trick is simply to duplicate each transistor, along with its gate terminal, and modify the current functions of each instance so that one is zero when $v_D > v_S$ and the the other is zero when $v_D < v_S$. No correlation is ignored in this transformation because only one instance of the gate voltage affects the circuit at a time, but the terminal currents are monotonic in each with a fixed polarity.

Once the gate terminals are separated from the other terminals, the monotonic property derived by assuming fixed gate voltages can still be used. The proper gate voltages must be used in each instance to form the extreme terminal currents. The magnitude of current produced by a one-port group is a monotonically increasing function of all its gate voltages, so the effect of a gate voltage on a particular terminal current of $\mathcal{N}_N$ depends on the reference direction of its one-port group with respect to the terminal. To keep the effects of gate terminals separate from each other and any source and drain terminals they are connected to, an extended input voltage vector $\mathbf{V}$ is defined, with $r-n$ new components corresponding to half that many gate terminals. Each of the new components corresponds to one of the original ones. For a terminal that was not connected to a source or drain in the original circuit, a dummy component which has no effect on the circuit is left in the original portion of the voltage vector for uniformity. The terminal constraints are generalized to the form $\mathbf{i} = \mathbf{F}^N(\mathbf{V})$.

The effect of transistors on the terminal constraints of $\mathcal{N}_N$ is almost completely analogous to that for $\mathcal{N}_R$. Each terminal current is a current-wise monotonically increasing function of each one-port group whose drain is connected to its terminal, and a current-wise monotonically decreasing function of any one-ports whose source is connected to its terminal. Each one-port group is then a monotonically increasing function of its internal *N*-elements, but unlike the resistor subnetwork $\mathcal{N}_R$, this monotonicity holds in either partial

---

[40] It is easy to show that the polarity of each transistor within a one-port group is always the same as that of the entire group.

[41] The only purpose of this circuit transformation is to simplify analysis. As with other transformations mentioned here, it does not imply that transistors must actually be duplicated in the database of a simulation program.

ordering for general N-elements[42].

The final issue that needs to be considered is the effect of the substrate voltage on the terminal constraints. If the substrate voltage is not known precisely, a relationship between the substrate voltage and other terminal currents must be determined. Fortunately, the substrate terminal of each transistor can be handled much as a gate terminal, since the substrate voltage has a similar monotonic effect on the current magnitude. The only difference is that the polarity is reversed. The substrate terminals are just treated like negative gate terminals. All substrate terminals (of n-channel transistors) are usually connected to a single external input terminal, called VBB, that is fixed at ground or some negative voltage.

4.3.4 Transistor Polarity

The subnetwork $\mathcal{N}_P$ is considered to extend the MOS circuit model to include CMOS circuits. A corresponding analysis of $\mathcal{N}_P$ need not be undertaken since, aside from polarity, it is identical to the analysis for $\mathcal{N}_N$. The analysis in this chapter is at a level of generality that is beyond particular threshold voltage magnitudes and carrier mobilities, in which n-channel and p-channel transistors do differ. Assuming all transistors are guaranteed to be biased for normal operation, the only differences between N-elements and P-elements are the polarities of the effects of gate and substrate voltages. With the appropriate changes in sign and directions of monotonic properties, all the analogous definitions and theorems for p-channel subnetworks are valid.

4.3.5 Transistor Subnetwork Summary

The results of this section are summarized in a definition and a theorem. The definition states the restrictions placed on $\mathcal{N}_N$ and the theorem states all of its important properties. The subnetwork $\mathcal{N}_P$ is completely analogous.

> **Definition 4-12:** The transistor subnetwork $\mathcal{N}_N$ is a multi-terminal resistor constructed from N-elements. Each internal element is part of a series-parallel one-port group (defined by ignoring gate and

---

[42] This would also be true for $\mathcal{N}_R$ if its one-port groups were required to be series-parallel.

substrate terminals) that satisfies a global Lipschitz condition and whose drain and source are external terminals (j for some $0 \leq j \leq n-1$) of $\mathcal{N}_N$. Each transistor gate and substrate terminal, except for gate terminals connected to the source or drain of the same transistor, is connected only to two external terminals (j for some $n \leq j \leq r-1$), one for each transistor polarity (depending on the sign of $v_D - v_S$ for its transistor), with no other transistor connections. Each gate and substrate terminal is a duplicated instance of one of the terminals numbered 0 to $n-1$, having identical voltage bounds. No terminal j for $j \geq n$ can be a duplicated instance of a terminal connected to the drain or source of the one-port that contains its gate terminal.

Theorem 4-5: The transistor subnetwork $\mathcal{N}_N$ defines a unique mapping $\mathbf{i} = F^N(\mathbf{V})$ satisfying a Lipschitz condition. Any component of $\mathbf{i}$, i_j for $0 \leq j \leq n-1$, satisfies the following: 1) it is a monotonically increasing function of V_j, 2) it is a monotonically decreasing function of V_k for $k \neq j$, $0 \leq k \leq n-1$, 3) it is a monotonically increasing function of V_k, $n \leq k \leq n+r-1$ if the one-port group associated with V_k has its drain connected to terminal j and V_k is associated with positive drain current or has its source connected to terminal j and V_k is associated with negative drain current, 4) it is a monotonically decreasing function of V_k, $n \leq k \leq n+r-1$ if the one-port group associated with V_k has its drain connected to terminal j and V_k is associated with negative drain current or has its source connected to terminal j and V_k is associated with positive drain current, 5) it is a constant function of V_k, $n \leq k \leq n+r-1$ if the one-port group associated with V_k has neither its drain or source connected to terminal j, 6) it is a current-wise monotonically increasing function of any N-element in a one-port group whose drain is connected to terminal j, 7) it is a current-wise monotonically decreasing function of any N-element in a one-port group whose source is connected to terminal j, and 8) it is a constant function of any other N-element.

Proof: By KCL, the current i_j for $0 \leq j \leq n-1$ is the sum of all drain currents from one-ports whose drain is connected to terminal j, minus the sum of all drain currents from one-ports whose source is connected to terminal j. Each of these components is the current of a series-parallel one-port group, so the whole function $F^R(V)$ is well defined, unique, and satisfies a Lipschitz condition by addition of functions with these properties. Monotonicity in the terminal voltages follows from addition and the results of theorem 4-4. Monotonicity in the N-elements follows from addition, theorem 4-4, and the definition of a current-wise ordering for N-elements. ∎

In order to produce simple monotonic relationships among the terminal variables and internal elements of the transistor subnetworks $\mathcal{N}_N$ and $\mathcal{N}_P$, their elements must be grouped into series-parallel one-port groups that directly connect external terminals. Using a range of device characteristics for each of the transistors along with bounds on the terminal voltages, efficient bounds on the terminal currents can be produced. An upper bound on a terminal current is calculated by summing the maximum currents that can leave through each series-parallel one-port group connected to the terminal. Each of these currents is calculated using the maximum voltage drop across the group, based on bounds for the terminal voltages. If the maximum current flowing through an n-channel group away from the terminal is positive, lower bounds on its transistors and gate terminal voltages are used along with an upper bound on its substrate voltage to transform the voltage drop into a current. Otherwise, the opposite extremes are used.

A transistor, being essentially a time-varying resistor, can produce a subnetwork that is difficult to evaluate efficiently. If complex series-parallel groups must be evaluated exactly for a large number of different gate voltage combinations, large amounts of computation can be required even when using simplified transistor bounds. Fortunately, the monotonic relationships presented in this section can often be used to simplify this calculation, at the expense of some accuracy, to that of finding the operating point of a single transistor. The details of such a transformation are considered in chapter six.

4.4 The Capacitor Subnetwork

This section considers the subnetwork of the MOS circuit model $\mathcal{N}_C$ containing all two-terminal capacitors. As with the other subnetworks, restrictions on the subnetwork that guarantee basic monotonic properties must be derived. In addition, restrictions must be considered that allow the application of relaxation techniques to the solution of the entire network. In most MOS circuit models, the capacitor subnetwork contains no internal nodes, i.e., all capacitors are connected at both ends to external terminals. As with the resistor subnetwork, though, the model can be generalized to include capacitors that are one-port capacitor groups. To allow the use of relaxation techniques based on simple nodal analysis, any pair of external terminals of $\mathcal{N}_C$ must be connected through a path passing only through capacitors. The topology of the capacitor subnetwork also influences the partitioning for relaxation. In order to guarantee the monotonic properties needed later, no block in a partition can be so large that is contains a capacitor loop, unless the capacitor subnetwork $\mathcal{N}_C$ is linear.

4.4.1 Evaluation of Circuit Dynamics

The capacitor subnetwork $\mathcal{N}_C$, as pictured in figure 4-20, affects the rest of the circuit through its terminal constraint $\mathbf{q}=\mathbf{f}^C(\mathbf{v})$. Since current is the time derivative of charge, this can be expressed alternately in terms of the incremental capacitance as $\mathbf{i}=\mathbf{C}(\mathbf{v})\dot{\mathbf{v}}$ where $\mathbf{C}(\mathbf{v})=[\mathbf{J}(\mathbf{f})]_{(\mathbf{v})}$. The vectors $\mathbf{q}$, $\mathbf{v}$, $\mathbf{i}$, and $\dot{\mathbf{v}}$ have as components the terminal variables for each of the terminals of $\mathcal{N}_C$. The capacitor subnetwork contains all of the memory, and therefore state information in the network.

As discussed later, some of the terminal voltages of $\mathcal{N}_C$ must be treated as dependent variables, so some notation must be introduced that divides the terminal variable vectors into two pieces. The first piece, given a subscript of 1, must include at least the input components 0 to m−1, and possibly some of the independent node components m to n−1. The second piece, given a subscript of 2, represents all remaining terminals. The terminal constraints can also be expressed as:

$$\begin{bmatrix} \mathbf{q}_1 \\ \mathbf{v}_2 \end{bmatrix} = \mathbf{h}^C\left(\begin{bmatrix} \mathbf{v}_1 \\ \mathbf{q}_2 \end{bmatrix}\right)$$

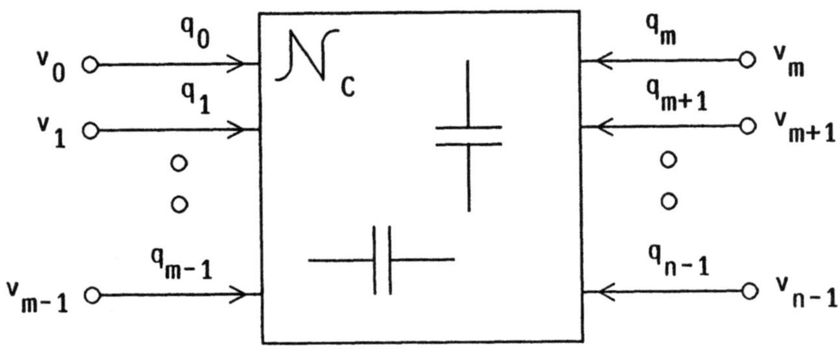

Figure 4-20: The capacitor subnetwork $\mathcal{N}_C$

$$\text{or} \quad \begin{bmatrix} \dot{i}_1 \\ v_2 \end{bmatrix} = g^C\left(v, \begin{bmatrix} \dot{v}_1 \\ i_2 \end{bmatrix}\right).$$

For exact analysis, the incremental terminal constraint has a simple linear form. When considering bounds, as will be seen later, it is often necessary to generalize this form, as done in the previous equation with the function g^C. Roughly speaking, this allows bounds on incremental capacitors to have different values depending on the polarity of their incremental current.

The network constraints are described using nodal analysis, as it provides an efficient set of equations for most digital MOS integrated circuit models. First, the method is based on using node voltages as independent variables. This is especially important for use with MOS devices. Second, the method produces a compact set of equations in a straightforward manner, using only the minimum state variables as unknowns. The basic approach is to generate one equation for each of the independent nodes (nodes m to $n-1$), based on KCL and expressed in terms of the state variables. A convenient choice for state variables is either[43] the terminal charges **q** or the terminal voltages **v** of $\mathcal{N}_C$.

If the function $\mathbf{h}^C$ is a well defined function, i.e., the dependent and independent variables can be inverted at any independent nodes, nodal analysis

---

[43] It is also possible to consider a mix of charges and voltages, but that possibility is not explored here.

produces a set of $n-m$ independent differential equations from the terminal constraints of each subnetwork. Invertibility implies that either $\{q_m, .. q_{n-1}\}$ or $\{v_m, .. v_{n-1}\}$ are equivalent to the state of the network, i.e., the state has $n-m$ dimensions. Without this restriction, the dimension of the state is less than $n-m$ and cannot completely determine, along with the inputs, the independent variables of the resistive subnetworks. Some of the network constraints then become awkward algebraic equations that complicate analysis. Therefore, existence and uniqueness of $\mathbf{h}^C$ is required.

The assumption that $\mathbf{h}^C$ is well defined has implications for the structure of $\mathcal{N}_C$ that may not be immediately obvious. It implies that each independent terminal of $\mathcal{N}_C$ is in fact connected through a finite, nonzero capacitive path to ground. This constraint only involves the addition of arbitrarily small capacitors to produce equations with a simple structure. The three resistive subnetworks can only be connected to each other at nodes that are also connected to capacitors. As a result, each of the terminal voltages for the resistive subnetworks $\mathcal{N}_R$, $\mathcal{N}_N$, and $\mathcal{N}_P$ are completely determined by the state of the network. The preponderance of capacitors connected to a common ground terminal in integrated circuits makes this invertibility constraint very mild in practice.

The resistive subnetworks $\mathcal{N}_R$, $\mathcal{N}_N$, and $\mathcal{N}_P$ each contribute currents to the nodes m to $n-1$ that are functions of the node voltages. These currents can simply be added to generate a total current contribution from the three resistive subnetworks. If charge is used as a state variable, then the total current represents the time derivative of the state. If the voltage is used as a state variable, then the total current is used as a terminal current for $\mathcal{N}_C$ to calculate the voltage derivatives. Figure 4-21 contains a dataflow diagram for the calculations needed to determine state trajectories, and therefore circuit behavior, for both choices of state variables. The two sets of equations are as follows:

$$C_i(v)\dot{v} = f_i^R(v) + f_i^N(v) + f_i^P(v) \quad , \quad m \le i \le n-1 \tag{4.1}$$

$$\frac{d}{dt} f_i^C(v) = f_i^R(v) + f_i^N(v) + f_i^P(v) \quad , \quad m \le i \le n-1. \tag{4.2}$$

Since nodal analysis is used to form the network equations, currents and charges become the convenient dependent variables. The dataflow diagram of

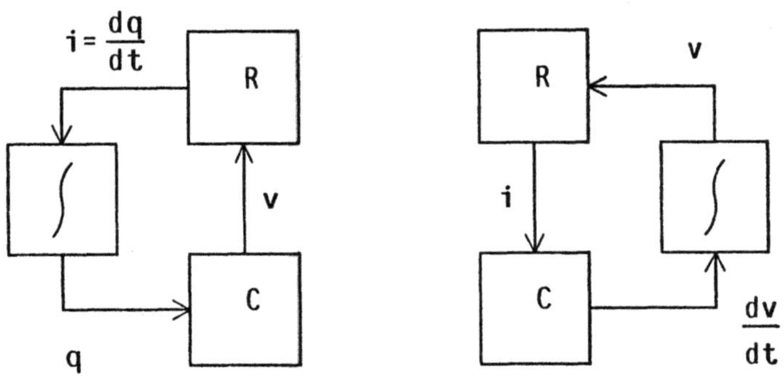

Figure 4-21: Dataflow diagrams for evaluation of node equations.

figure 4-21 shows, however, that the capacitor subnetwork $\mathcal{N}_C$ must be evaluated "backwards," i.e., the constitutive relation must be partially inverted, to obtain the time derivative of the state. Since a cycle must be formed in this diagram, switching to loop analysis only shifts the problem to the resistive subnetworks. Due to the locality of connections in an integrated circuit, the constitutive relation of the resistive subnetworks can generally be calculated locally, i.e., each current depends only on a few voltages, and with little difficulty. In general, due to internodal coupling capacitance and the "backwards" calculation of voltages or their derivatives, the constitutive relation of the capacitive subnetwork is not local and is difficult to calculate. In the special case of a completely linear network, a sparse capacitance matrix must be inverted, producing a matrix that is no longer sparse. Although the inverted matrix is not sparse, most terms are relatively small compared to the diagonal terms. Each terminal voltage of $\mathcal{N}_C$ is only a strong function of local terminal charges. This observation is promising with regard to the application of relaxation methods for calculation of the terminal voltages.

4.4.2 Waveform Relaxation and the Capacitor Subnetwork

In this subsection the convergence of exact waveform relaxation algorithms is considered. As shown in detail in the next chapter, many properties of bound relaxation depend on the convergence property of an underlying exact algorithm, even though the exact algorithm is not explicitly evaluated. An intuitive explanation of Waveform Relaxation is presented, followed by a demonstration that any partitioning of the node equations (equations (4.1) or (4.2)), with each variable assigned to its corresponding node equation, is valid under mild conditions. Roughly speaking, this assignment means that node equation i is used to update guesses for the behavior of v_i. More complete discussions of Waveform Relaxation can be found in [11] and [48].

A waveform relaxation algorithm conceptually involves two distinct types of relaxation. The first involves a relaxation over time. Consider taking a guess for the state space trajectory ($x^0(t)$ for $0 \leq t \leq T$) of the solution to a vector differential equation of the form $\dot{x} = f(x)$ where $x(0) = x_0$. Pick the guess so that its initial value is correct, but the behavior $x^0(t)$ might be completely wrong. One reasonable way to improve the guess would be to start at the initial point and calculate a new guess using at each value of time the state derivative given by the state velocity vectors along the old trajectory. The second guess:

$$x^1(t) = x_0 + \int_0^t x^0(t)dt$$

is guaranteed to at least have the correct derivative at $t=0$, so the guess is good for small times. If the state velocity vector function $f(x)$ satisfies simple continuity constraints, repeating the process of improving guesses eventually leads arbitrarily close to the solution. Due to the continuity requirement, a good solution for small times produces a good guess for the solution a little later, leading eventually to a good solution for the entire time interval. This process is known as Picard iteration [49].

The second type of relaxation takes place in space. The space can be viewed in terms of the physical circuit, or in terms of the state space. Any subvector of the state vector represents the behavior of a subcircuit. Consider a new algorithm for updating a guess for the trajectory. Assume that the vector $\mathbf{x}$ is partitioned into subvectors $\mathbf{x}_i$, and each subvector trajectory $\mathbf{x}_i^k(\cdot)$ is calculated separately over [0,T], using the old trajectory $x^{k-1}(\cdot)$ only to provide information about components of $\mathbf{x}$ missing from $\mathbf{x}_i$. If coupling is weak between these groups, this strategy would appear to improve guesses more quickly than the Picard iteration described above. Differential equations must be solved but the partitioning discussed in section 2.2 has been partially accomplished. Small subcircuits are analyzed separately, but components of the velocity vector might still be a function of the entire state. This algorithm is guaranteed to converge for the same reasons as before. The second guess still obtains a correct first derivative, albeit one group at a time, and the same rough inductive argument extends the quality of the guesses eventually throughout the entire time interval.

The next logical question is whether the space partitioning can be used additionally to calculate the state velocity vectors. As hinted at the end of the last subsection, the calculation of the state velocity vector, represented by the dataflow diagrams in figure 4-21, could also benefit from relaxation in space. Due to the capacitor subnetwork, a single component of the state velocity vector is generally a function of all others and the partitioning discussed in section 2.2 is not complete. We desire calculations involving small subcircuits with only local influences to obtain linear scaling of computation with circuit size. Fortunately, it has been shown that this additional space partitioning can be used [11]. Roughly speaking, if the calculation used to update a guess for a state velocity vector by components is contractive, the sequence eventually achieves an arbitrarily good guess for the derivative at $t=0$. Continuity requirements can still be used to extend this convergence inductively to the entire interval. The Waveform Relaxation proof has recently been redone in a form that more clearly distinguishes relaxation of the state velocity vector as the key component [48].

The Waveform Relaxation theorem is formulated with the inclusion of algebraic equations in each component of the space partition. This extends the contraction property required in the calculation of the state velocity vector to include the algebraic equations. One component of the calculation is represented by the vector equations (4.3). In equation (4.3) the superscripts denote an iteration count. A component of the state velocity vector $\dot{x}_i^k$, and its associated algebraic variables z_i^k, are calculated from the corresponding component of the state, as well as old guesses for all variables (including both state and state velocity components) associated with other components.

$$\dot{x}_i^k = f_i(x^k, x^{k-1}, \dot{x}^{k-1}, z^{k-1}), \qquad z_i^k = g_i(x^k, x^{k-1}, \dot{x}^{k-1}, z^{k-1}). \tag{4.3}$$

To evaluate the constitutive relation of $\mathcal{N}_C$ in a mostly forward direction, i.e., use mostly voltages or voltage derivatives as independent variables, old guesses are used to supply terminal voltages or derivatives from other components of the circuit. The calculation of v or $\dot{v}$ in each component then becomes a local calculation. When voltage is used as a state variable, the relaxation at each time involves the voltage derivatives, corresponding to x in eqn. (4.3). When charge is the state variable, the voltage is an algebraic quantity corresponding to z in equation (4.3). A single component of equations (4.1) or (4.2) expressed in the form of equation (4.3) with only local calculations[44] becomes, respectively:

$$\dot{v}_i^k = f_i(v^k, v^{k-1}, \dot{v}^{k-1}) \tag{4.4}$$

$$\dot{q}_i^k = f_i(v^{k-1}), \qquad v_i^k = g_i(q^k, v^{k-1}). \tag{4.5}$$

For a waveform relaxation algorithm to converge, the mapping from one set of voltages or voltage derivatives to another, for a given state, must be contractive. In either equation (4.4) or (4.5), this condition is satisfied since every terminal node of $\mathcal{N}_C$ is connected through a path of nonzero capacitors to ground. The conditions for guaranteeing the convergence of a waveform relaxation algorithm are extended here to a wider class of networks than presented in [11]. Each node m to $n-1$ need not be connected to a grounded capacitor, only to a path of capacitors leading to ground as mentioned in the previous subsection. In addition, the equations formed when charge is used as a state variable can also be used in a waveform relaxation algorithm. The mappings in equations (4.4) and (4.5) only need to satisfy a Lipschitz condition, with respect to the state variables, that follows from the results in the last two sections, as well as results presented later in this section. The Lipschitz condition implies the existence and uniqueness of a solution for each subcircuit [49], and the convergence of a waveform relaxation algorithm implies existence and uniqueness of a solution for the entire network.

To extend the convergence of waveform relaxation algorithms to cover partitionings of both equations (4.1) and (4.2), contraction of the mappings leading to the new state velocity vectors at each value of time must be shown. The state variables are treated as constants in this mapping. In the first case the mapping is expressed as $\dot{v}^k = \hat{\beta}(\dot{v}^{k-1})$ and in the second it becomes

[44] Equations (4.1) and (4.2) can be solved for each state derivative component as a function only of the state, producing a set of equations said to be in "normal" form. In general, each component of the derivative is a function of each state component in normal form so the equations are no longer local. By maintaining the form of equations (4.1) and (4.2), each component can be solved for its corresponding state derivative component with only information from nodes directly connected through circuit elements.

$v^k=\beta(v^{k-1})$. In each case the mapping is defined by components. New voltages or voltage derivatives are calculated for each group of terminals by solving for the operating point of $\mathcal{N}_C$ with the other terminal voltages fixed at their previous guesses[45] (v^{k-1} or $\dot{v}^{k-1}$) and the terminal charges or currents for the group fixed at their known values determined by the state[46]. For the function β, this is:

$$v_i^k = \beta_i(v^{k-1})$$
$$= h_i^C([v_0^{k-1}, .. v_{i-1}^{k-1}, q_i, v_{i+1}^{k-1}, .. v_{n-1}^{k-1}]^T) \text{ for } m \leq i \leq n-1.$$

Recall that the object of the relaxation is to solve for the correct terminal voltages (or voltage derivatives), given a set of terminal charges (or currents) as independent variables. The solution is a fixed point of the relaxation function $\hat{\beta}$ (or β).

The contraction property is demonstrated here for the more general relaxation function β, since $\hat{\beta}$ is just a special case analogous to evaluating β in a circuit with only linear capacitors. The function β is not actually strictly contractive itself, but a finite number of compositions of β with itself produces one that is. This still implies the convergence of a waveform relaxation algorithm simply by a corresponding redefinition of the relaxation function. The norm used on the spaces of terminal voltages is simply the maximum voltage magnitude. Note that based on definition 4-5, along with the fact that there is a finite number of capacitors in $\mathcal{N}_C$, all capacitors must be incrementally bounded by two linear capacitors C_{min} and C_{max} such that $C_{min}>0$ and $C_{max}<\infty$. Since these incremental bounds are valid for all voltages, the capacitors C_{min} and C_{max} are simple bounds as well.

---

[45]The Gauss-Jacobi form of relaxation is used here as a simplification, but similar results hold for the Gauss-Seidel form as well, with only slight modifications of the arguments.

[46]The well-defined nature of this mapping can be easily shown when only one terminal voltage is unknown in each calculation, as there must be a capacitor connected to each node with the property that $q\rightarrow\infty$ when $v\rightarrow\infty$. Theorem 4-7 can then be stated first in terms of partitions containing only one node and combined with the contraction mapping theorem [50] to show a unique solution exists for larger partitions even with direct capacitive coupling between the nodes in the partition. An alternative approach is presented in theorem 4-8.

Proving contraction of the mapping β starts with a simple lemma on bounding the voltage of the internal node in a series capacitor group. We assume that all internal nodes in $\mathcal{N}_C$ have zero net charge, i.e., the charge analogy of KCL is valid, so $\mathcal{N}_C$ is analogous to $\mathcal{N}_R$ if charges are replaced by currents. The analogous result for resistor subnetworks can be used to bound the internal behavior of the resistor subnetwork when only series-parallel groups are used.

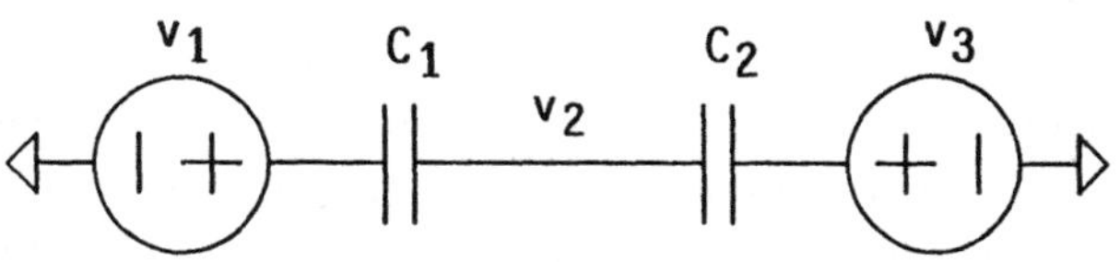

Figure 4-22: Series combination of capacitors.

Lemma 4-6: Consider the capacitor circuit in figure 4-22. If $v_1 \geq v_3$, v_2 is a monotonically increasing function of C_1 and a monotonically decreasing function of C_2.

Proof: Assume C_1 is replaced by a larger capacitor and v_2 decreases. Charge must enter node two from C_2 since v_2 decreased and the incremental capacitance of C_2 is positive. The increase in C_1 and decrease in v_2 cause charge to enter node two from C_1 as well, leading to a contradiction of conservation of charge on node two. A similar contradiction arises if C_2 is replaced by a larger capacitor and v_2 increases. The result then follows from these two contradictions. ∎

Theorem 4-7: Let β be the mapping $\mathbb{R}^{n-m-1} \rightarrow \mathbb{R}^{n-m-1}$ previously defined. Let $\beta^{(p)}$ represent the function β composed with itself p times. Let $\mathbf{v}$ and $\hat{\mathbf{v}}$ be any two terminal voltage vectors. Then there exists some integer $p>0$ and some $\alpha \in (0,1)$, both independent of $\mathbf{v}$, $\hat{\mathbf{v}}$, and the terminal charges used in the definition of the function $\boldsymbol{\beta}$, such that $\|\beta^{(p)}(\mathbf{v}) - \beta^{(p)}(\hat{\mathbf{v}})\| < \alpha \|\mathbf{v} - \hat{\mathbf{v}}\|$.

Proof: Pick any terminal voltage vectors $\mathbf{v}$ and $\hat{\mathbf{v}}$ and let $v_{max} \stackrel{\Delta}{=} \|\mathbf{v} - \hat{\mathbf{v}}\|$. Consider the sequence $\{\mathbf{v} - \hat{\mathbf{v}},\ \beta(\mathbf{v}) - \beta(\hat{\mathbf{v}}),\ \beta^{(2)}(\mathbf{v}) - \beta^{(2)}(\hat{\mathbf{v}}),\ \ldots\}$ where each element is obtained from the previous one by solving an incremental (not linearized) circuit. Each charge source becomes disconnected in the incremental circuit so the sequence is independent of the terminal charges. Note that by applying the voltage minimax theorem (see proof of theorem 4-1) to capacitor circuits, when the incremental input is $\mathbf{v} - \hat{\mathbf{v}}$, all incremental node voltages must fall in the interval $[-v_{max}, v_{max}]$. Furthermore, an incremental application of the voltage minimax theorem implies that each internal node of a capacitor network has a voltage that is monotonic in its terminal voltages. Let γ be the number of capacitors in $\mathcal{N}_C$. When this is combined with the result of lemma 4-6, it follows that any internal node (including those with disconnected charge sources) connected through a capacitor to a node with an incremental voltage contained in $[-v_*, v_*]$ for some $0 < v_* < v_{max}$, has an incremental voltage bounded by $[-(v_* + (v_{max} - v_*)(\gamma C_{max}/(C_{min} + \gamma C_{max}))),\ v_* + (v_{max} - v_*)(\gamma C_{max}/(C_{min} + \gamma C_{max}))] \subset (-v_{max}, v_{max})$. Some terminal other than ground, denoted i, must be connected to ground through a path that does not traverse any other terminals, or the path constraint to ground is violated. Since the incremental voltage at the ground terminal is contained in $(-v_{max}, v_{max})$, v_i is also by induction along the path. Therefore, $|\beta_i^{(1)}(\mathbf{v}) - \beta_i^{(1)}(\hat{\mathbf{v}})| < v_{max}$. Another terminal j must be connected to ground or terminal i through a path that does not traverse other terminals, and by the same argument $|\beta_j^{(2)}(\mathbf{v}) - \beta_j^{(2)}(\hat{\mathbf{v}})| < v_{max}$. $|\beta_i^{(2)}(\mathbf{v}) - \beta_i^{(2)}(\hat{\mathbf{v}})| < v_{max}$ as before. Eventually all terminals must be exhausted in this manner after some finite number of iterations $p \leq \gamma$ corresponding to the longest capacitor path leading to ground. Therefore, $\|\beta^{(p)}(\mathbf{v})-\beta^{(p)}(\hat{\mathbf{v}})\|<v_{max}$ since the number of dimensions is finite. Some $\alpha<1$ can then be chosen to satisfy $\|\beta^{(p)}(\mathbf{v})-\beta^{(p)}(\hat{\mathbf{v}})\| < \alpha\|\mathbf{v}-\hat{\mathbf{v}}\|$ for the particular sets of voltages $\mathbf{v}$ and $\hat{\mathbf{v}}$. The linear bounds used are valid for any such voltages so the final result follows. ∎

4.4.3 Terminal Constraints

If charge is used for a state variable, the analysis of the capacitor subnetwork is completely analogous to that of the resistor subnetwork. If the restriction that there can be no net charge at any internal node of $\mathcal{N}_C$ is added, the capacitor case is identical since current in the analysis of $\mathcal{N}_R$ can be replaced everywhere by charge. The constraint of zero net charge on internal nodes then takes the place of KCL in generating the same governing equations. In addition, the definition of a capacitor is a special case of that of the resistor, if they are compared by replacing charges with currents in the capacitor definition. As a result, there is a unique mapping $\mathbf{q}=\mathbf{f}^C(\mathbf{v})$ which is monotonic as defined for $\mathcal{N}_R$, i.e., each terminal charge is a monotonically increasing function of its terminal voltage and a monotonically decreasing function of all other terminal voltages. When determining the operating point of $\mathcal{N}_C$ for incremental analysis it should be noted that all the internal node voltages are monotonic functions of the terminal voltages by analogy with the proof of theorem 4-1.

When considering the terminal constraints for an incremental ordering, it must be assumed that the incremental capacitance matrix has been evaluated at a particular operating point, determined by the terminal voltages or charges. Based on the monotonic properties just discussed for $\mathbf{f}^C(\mathbf{v})$, the diagonal terms of the matrix $\mathbf{i}=\mathbf{C}(\mathbf{v})\dot{\mathbf{v}}$ are nonnegative and all others are nonpositive. The subnetwork $\mathcal{N}_C$ has the same monotonic terminal constraints in the incremental case as it does in general. The incremental subnetwork is a special case of the complete one in which all the elements are guaranteed to be linear, at least in the exact case.

When calculating tight bounds on incremental terminal constraints, the operating point must first be considered. To transform ranges of node voltages to ranges of incremental capacitance for each element, it helps if the incremental capacitance is a monotonic function of voltage. A reasonable restriction is that the incremental capacitance of each element is either a nonincreasing or nondecreasing function of voltage. In this case only two endpoints need be evaluated for each element. This is not a requirement, though. It is assumed that the operating point is treated as a constant, i.e., has constant bounds, over significant time intervals. A modest amount of computation can then be tolerated in converting these operating point bounds into incremental capacitor bounds that are valid throughout the time interval.

For either set of terminal constraints, consideration must be given to the fact that $\mathcal{N}_C$ is not analyzed in voltage-controlled form. For some subset of the terminals, charges are the independent variables and voltages are calculated. Due to the constraint that all capacitors must have positive incremental capacitance, each terminal constraint has a property analogous to incremental passivity in $\mathcal{N}_R$. As a result, any change that increases the charge for any given voltage must decrease the voltage for any given charge. Also, if the charge at a terminal is increased, the terminal voltage must also increase, having the same effect. The dependent and independent variables can be switched at some terminals and the terminal constraints maintain their monotonicity.

Consider the situation pictured in figure 4-23 where $\mathcal{N}_C$ is driven by both charge and voltage sources. Any topology is allowed for $\mathcal{N}_C$ at this point as long as there is a path through capacitors between any independent node and ground. Inputs and old guesses for independent node voltages are labelled v_1 to v_{p-1}, and the unknown node voltages are labelled v_p to v_{n-1} for p>m. No generality is lost by renumbering the nodes for analysis so that all charge sources appear on the terminals with the largest numbers. The following theorem considers the important general properties of $\mathcal{N}_C$ that do not require additional topology constraints. It is stated in terms of charges and voltages, but the incremental case with currents and voltage derivatives is a special case.

Theorem 4-8: The circuit pictured in figure 4-23 defines a unique set of voltages v_i for $p \leq i < n$ that are Lipschitz in each input source (this is equivalent to saying the complete mapping function is Lipschitz). Furthermore, v_i for any $p \leq i < n$ is a monotonically increasing function of all the charge sources and all the voltage sources.

Proof: A unique set of branch voltages exists for any network such as the one in figure 4-23 by analogy to resistor networks ([51], p. 776), and since there is a capacitor path between every independent terminal and ground, this uniquely determines the voltages v_i for $p \le i < n$. Lipschitz continuity with respect to the voltage sources follows from an incremental application of the voltage minimax theorem (the charge sources are incremental open circuits). The incremental voltage gain of the circuit must be less than unity. For Lipschitz continuity in the charge sources, first note that the charge magnitude in any element of a one-port is bounded by that of the port charge ([51], p. 780). Consider the incremental circuit formed by increasing one of the charge sources by an amount Δq. The magnitude of incremental charge on each element is bounded by Δq, and since each element constraint is Lipschitz, the incremental voltage magnitudes are bounded by $k\Delta q$ for some k. Call the number of capacitors in the circuit r. Since any node is connected through a path of capacitors to ground, its incremental voltage change is bounded by $rk\Delta q$. This implies Lipschitz continuity of each component in each source. For monotonicity in the voltage sources again consider an incremental application of the voltage minimax theorem. Any positive increment in a voltage source cannot produce a negative increment in an output voltage. For a charge source, a positive charge increment must produce a nonnegative voltage increment at its terminal by analogy to incremental passivity of resistor networks. As with the voltage sources, this nonnegative increment in voltage cannot lead to a negative increment in any other output voltage either. ∎

4.4.4 Internal Elements

When considering the effects of internal elements on the terminal constraints, the capacitor subnetwork $\mathcal{N}_C$ can again be made completely analogous to $\mathcal{N}_R$. The use of some terminal charges as independent variables, though, makes a more thorough analysis of the effects necessary. At least for operating point calculations, the terminal voltages are given, so the same topology constraints imposed on the resistor networks are used here. The one-port grouping constraint is not a fundamental restriction on the types of circuits that can be considered, even for the capacitor subnetwork. In the unlikely event that there is an internal node in $\mathcal{N}_C$ that violates the constraint, it can be connected to an arbitrarily large grounded resistor. Such a change of circuit model is analogous to adding small capacitors when considering the resistor subnetwork.

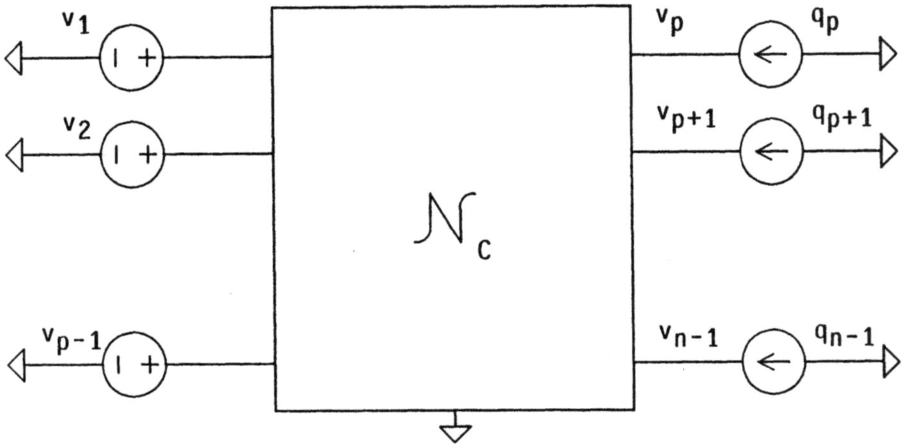

Figure 4-23: A number of terminal voltages can be solved at once.

If the operating point of $\mathcal{N}_C$ is needed to calculate an incremental terminal constraint, the effect of the elements on the internal nodes must also be considered. In this case a further restriction is useful. If the one-port groups are required to be entirely series-parallel, whether linear or not, the internal node voltage ranges can be easily determined. Initially, one-port groups with known terminal voltage ranges exist. Each one-port is then divided recursively into its series-parallel components based on definition 4-8 applied to capacitors. For any parallel combination, no new internal nodes are produced. For a series combination, the internal node voltage is a monotonically increasing function of the two terminal voltages. It is also a monotonically increasing function of the capacitor connecting it to the terminal with the lower voltage, and a monotonically decreasing function of the capacitor connecting it to the terminal with the higher voltage. These monotonic relationships follow from lemma 4-6.

To find an upper bound on a node voltage in a series-parallel group, an upper bound on the two terminal voltages must be used. The group is partitioned into its two subgroups. For parallel connections, one obtains two groups with identical terminal voltages, each of which must be analyzed as the entire group is. For series connections, an upper bound on the internal node voltage must be obtained to continue analysis of the two internal groups. This bound is obtained by calculating the lower bound on the subgroup connected to the larger terminal voltage, along with an upper bound on the other. These calculations use the extremes for each of the internal elements of the subgroups. The extreme voltage is then calculated with a simple voltage divider calculation.

Whether the analysis is incremental or not, the one-port group capacitors are monotonic in their internal elements by analogy with theorem 4-1. From this point on only the one-port group capacitors are considered. It is assumed that they can have internal elements and that any bounds on the one-port group are derived from those on its internal elements.

When some of the terminal voltages are treated as dependent variables, the monotonic properties of the subnetwork with respect to the elements are not as simple as those derived for the resistor subnetwork. For linear networks arising from linear capacitor models or the use of incremental analysis, the relationships are monotonic but the sign is a function of other elements. For the nonlinear case, restrictions must be placed on the partitions chosen to guarantee even monotonicity with either sign. Partitions corresponding to one or two nodes can always be used in non-incremental analysis, but there are some cases in which larger clusters cannot.

Consider first the case pictured in figure 4-24 where only one terminal voltage, v_3, is being calculated. In general the capacitors may not be linear, but it is still the case that the unknown voltage must be a monotonic function of each element, given values for the other elements. The monotonic relationship is demonstrated in theorem 4-9. Even with such monotonicity, it is not clear how to determine the set of extremes in elements to be used to calculate a bound on v_3. The only thing that can be immediately determined is that one of the sets of extremes does indeed produce a bound[47]. As a result, if all combinations are tried and the most extreme results are chosen, they must be valid bounds. In fact, these extremes are the best possible bounds because they are also valid solutions if each capacitor can attain either of its bounds.

Now assume that the circuit in figure 4-24 contains linear capacitors, possibly arising from incremental analysis (in which case voltage derivatives replace voltages), and v_3 is being calculated as a function of q_3 and the other terminal voltages. The equation for v_3 is

---

[47] Starting with each capacitor at its actual value and moving each to its proper extreme one at a time produces a valid bound corresponding to one set of extremes. This argument shows that one extreme is a bound, but since the actual values are not known during computation, it does not produce a procedure for finding it.

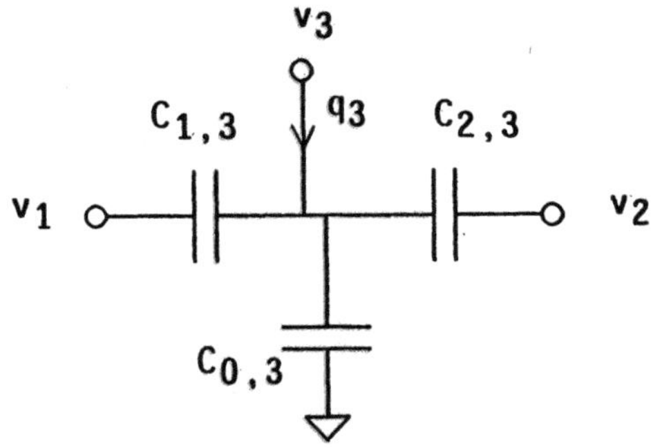

Figure 4-24: Bounding the effect of internal elements in $\mathcal{N}_C$.

$$v_3=(1/(C_{0,3}+C_{1,3}+C_{2,3}))(q_3+v_1C_{1,3}+v_2C_{2,3})$$

and v_3 is an increasing function of $C_{1,3}$ if $v_1(C_{0,3}+C_{2,3})>q_3+v_2C_{2,3}$. Notice that the condition does not depend on $C_{1,3}$ or v_3, so varying $C_{1,3}$ cannot change the polarity of voltage across it. The number of experiments needed is not usually a significant problem due to the small number of capacitors connected to any given node. For the linear case, a simplification can be made at the expense of some accuracy to avoid exponential searches. The numerator of the expression for v_3 is a monotonic function of each term. Therefore, to calculate an upper bound on v_3, an upper bound on the capacitance is used in the numerator if its coefficient is positive and a lower bound is used if the coefficient is negative. After an upper bound on the numerator has been found, a lower bound on the denominator is used only if the numerator is positive, otherwise an upper bound is used. The simplification ignores the correlation between multiple appearances of each capacitor value in the equation.

The linear simplification is equivalent to a circuit transformation where each capacitor is replaced by two capacitors and two dependent charge sources as pictured in figure 4-25. This transformation has been used in circuit simulation in the context of making approximations for the effects of floating capacitors [3]. The two capacitors C and $\hat{C}$ are identical for exact analysis, so making the transformation has no effect on the exact solution. For bounding analysis, the correlation between the two can be ignored to simplify computation at the expense of accuracy.

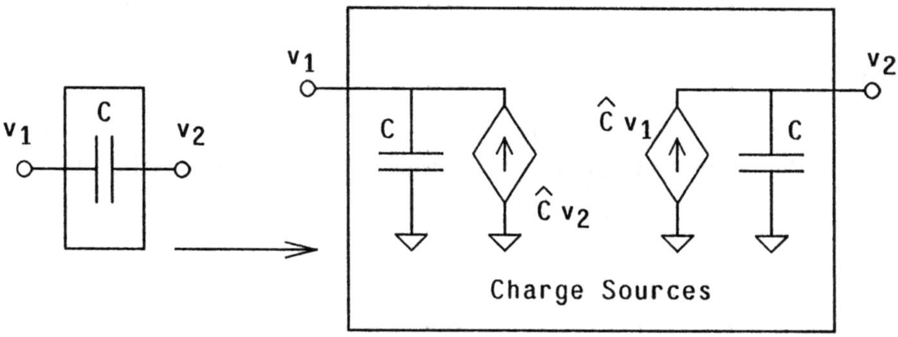

Figure 4-25: Circuit transformation that can simplify bounding of $\mathcal{N}_C$.

We now consider the conditions needed for element monotonicity. The general situation is pictured in figure 4-23. A circuit partition is being considered that includes nodes labelled p to $n-1$ where p>m. Inputs and old guesses for independent node voltages are being used for the voltages v_1 to v_{p-1}, and the voltages v_p to v_{n-1} are being calculated given the terminal charges q_p to q_{n-1}. Recall that the one-port capacitor groups are being considered as individual elements, so there are no internal nodes in $\mathcal{N}_C$ as pictured in the figure. For a nonlinear circuit, the capacitors that directly connect two of the nodes with fixed charge sources cannot form any loops. If they were allowed to, a nonmonotonic circuit analogous to the resistor circuit pictured in figure 4-11 could be produced. By setting some of the charge sources to zero, their terminals become essentially internal nodes.

The relationships between internal elements and terminal constraints are more easily stated in terms of charge-wise and current-wise orderings. A current-wise ordering for incremental capacitance, as formalized in the following definition, is analogous to a charge-wise ordering for simple capacitors. With these orderings, the direction of monotonicity for any element does not depend on either the terminal variables or the other elements. This does not really solve the problem of searching for the right extreme, however; it only disguises it. Charge-wise and current-wise orderings can only produce simplified elements that are piecewise linear, as discussed in section 4.1. For the resistor and transistor networks, the overhead of having breakpoints is low because, whenever the transfer function of the subnetwork is being evaluated, a comparison of two voltages determines the region of operation. Another way to look at the search problem in the capacitor subnetwork is that, because the calculations are "backwards," a search must be undertaken to find the region of operation.

> **Definition 4-13**: Given two *C*-elements C and $\hat{C}$ characterized by $i=g(v,\dot{v})$ and $i=\hat{g}(v,\dot{v})$ respectively[48], we say that " $\hat{C} \geq C$ current-wise" iff $\hat{g}(v,\dot{v}) \geq g(v,\dot{v})$ $\forall v$ and $\forall \dot{v}$.

The following theorem is stated in terms of charges, voltages, and charge-wise orderings, but applies to incremental analysis as a special case if the operating point is assumed to be fixed. The quantities become currents, voltage derivatives, and current-wise orderings. If the circuit is linear, even without the loop constraint there is a monotonic relationship between each capacitor and each terminal voltage derivative. This is a result of the fact that for a linear circuit the derivative of each relationship is independent of the operating point. There is no simple rule to determine the direction, when loops exist, but the search method can always be used.

> **Theorem 4-9:** Consider the capacitor subnetwork pictured in figure 4-23 with no internal nodes. Assume that the "floating" capacitors, those that directly connect two terminals with fixed charge sources, do not by themselves form any loops. Then v_i $\forall p \leq i < n$ is a charge-wise monotonic function of each capacitor connected to terminal i, where the function is increasing if the reference direction points towards the terminal and decreasing otherwise. Furthermore, v_i $\forall p \leq i < n$ is a charge-wise monotonic function of each capacitor that is connected to terminal i through a (unique) path of floating capacitors, where the polarity is increasing if the the reference direction points

---

[48] If the less general form $i=c(v)\dot{v}$ were used exclusively for incremental analysis, it would be impossible for two capacitors to be current-wise ordered.

towards the terminal along the path and decreasing otherwise. Finally, v_i $\forall p \leq i < n$ is not a function of the capacitors that do not fall in either class.

Proof: By analogy with incremental passivity in the resistor subnetwork, the subnetwork places a constraint on the charge and voltage of each capacitor such that the charge is a nonincreasing function of voltage. In addition, each capacitor has a charge that is an increasing function of voltage. As a result, a charge-wise increase in the capacitor is guaranteed not to decrease the charge on the capacitor or increase the voltage. Any capacitor not connected to a charge source has no effect on the voltages v_i for $p \leq i < n$. Let i be any terminal such that $p \leq i < n$. Consider replacing one capacitor connected to terminal i by a larger one in the charge-wise ordering, and without loss of generality, assume the reference direction is towards terminal i (switching direction just switches the polarity of all arguments). If the capacitor is not floating, the terminal voltage cannot increase because the capacitor voltage cannot increase. If the capacitor is floating (connecting terminal i with terminal j for $p \leq j < n$) and there is no other capacitor connected to terminal i, the charge on the capacitor must remain the same so no other node voltages change. Since the capacitor voltage cannot increase, the v_i cannot increase and as a result of theorem 4-8, neither can any terminal voltages. Terminals not connected through a path of floating capacitors cannot be affected as they are part of independent pieces. If the capacitor is floating and there is another capacitor connected to terminal i, but no other one connected to j, the current again is fixed and v_i cannot change. If the capacitor is floating and both terminals i and j are connected to other capacitors, replace the capacitor by a charge source with the same charge, both before and after the capacitor increase. Transform the circuit to an equivalent one where the charge source is removed, and its value is added to q_j and subtracted from q_i. The value of the charge source cannot decrease, so q_j cannot decrease and q_i cannot increase. Since floating capacitors form no loops, removing one divides the network into two independent pieces. The voltages of all terminals still connected to terminal i through a path of floating capacitors cannot increase as a result of theorem 4-8. All cases are then covered and the result follows. ∎

The loop constraint required for nonlinear networks does not affect the generality of the model, only the size of the partitions that can be treated efficiently. The strongly connected clusters that are often the most efficient partitions have nodes that are coupled through resistors, but not generally directly coupled through "floating" capacitors. As a result, the loop constraint is rarely a practical limit on partitioning.

4.4.5 Capacitor Subnetwork Summary

The relationships between voltages and charges in $\mathcal{N}_C$ are completely analogous to those between voltages and currents in $\mathcal{N}_R$. Incremental analysis, in which the terminal variables become voltage derivatives and currents, is a special case of the analogy with exclusively linear elements. In order to produce simple monotonic relationships among the terminal variables and internal elements of the capacitor subnetwork $\mathcal{N}_C$, the same one-port grouping constraint used for the resistor subnetwork is required. In addition, it's easier to calculate internal node voltages to determine the operating point for incremental analysis if the one-port groups are series-parallel.

The unique feature about $\mathcal{N}_C$ is that the terminal voltages are treated as dependent variables, and that relaxation can be used to make their calculation local. Mild continuity and topology constraints are required to allow the use of relaxation. Each terminal node in $\mathcal{N}_C$ must be connected to all others through some finite, nonzero capacitive path. This topology constraint is also sufficient to allow the resistive subnetworks to be evaluated with terminal voltages being completely independent variables. Nodal analysis is used to form the network equations.

Due to the partitioning of relaxation, calculations for the capacitor subnetwork are done with only a few voltages at a time being dependent variables. An upper bound on each voltage (or voltage derivative) is calculated using an upper bound for all the terminal charges (or currents) and an upper bound on all terminal voltages (or voltage derivatives). In general all combinations of extreme element constraints must be tried to determine the worst case. For linear or incremental circuits the search can be avoided by considering the equations directly and neglecting correlations between the numerator and denominator.

If $\mathcal{N}_C$ is analyzed in an incremental sense, the operating point must be computed first. If tight bounds on the incremental capacitance that depend on the operating point are desired, $\mathcal{N}_C$ must first be evaluated for each value of time (or more practically for a time interval) for all internal node voltages, with the terminal voltages being the independent variables. Each node voltage can be bounded by working down through each one-port group if the groups are guaranteed to be series-parallel. Bounds on the node voltages are then used to generate bounds on the incremental capacitance of each element.

In exact analysis, charge is sometimes preferred as a state variable due to numerical considerations. Incremental analysis produces additional error when numerical noise causes the linearization to become invalid. The more familiar quantity of voltage, though, is used in most MOS simulators. As illustrated in equations (4.1) and (4.2), the equations have a simpler form when voltage is used. Of course, the two methods are numerically equivalent if only linear capacitors are present in the network. In bounding analysis, some information is lost with either choice of state variable. In general, incremental bounds imply only loose simple bounds, and simple bounds do not place any restrictions on the incremental capacitance. Both choices produce *bounding simplification* which is different even if the capacitors are constrained to be linear. Due to the simple form of the equations, the simplifications with respect to the capacitor subnetwork calculations, and the familiarity of voltages, only voltages will be discussed in the remainder of the book. An investigation of the use of charge as a state variable requires only slight modifications from this point on.

4.5 Node Circuits

Now that all the subnetworks corresponding to different element types have been discussed, this section considers the properties of their interaction. This interaction is governed by the node equations that determine the behavior of the entire circuit. The monotonic properties of the solution for each independent node equation, corresponding to the most fine-grained partitioning of the state for relaxation, are derived.

4.5.1 Single Node Equation

Equation (4.1) in section 4.4 contains one possible set of differential equations that governs the behavior of the network $\mathcal{N}$. This set of equations contains $n-m$ separate node equations, one corresponding to each of the independent nodes at the boundaries of the element subnetworks. For relaxation algorithms, each of these node equations can be analyzed separately and solved for its corresponding node voltage. Each node equation corresponds to the solution of a subcircuit with only a single node and the elements (or one-port element groups) connecting it to input voltage sources, as pictured in figure 4-26. All other node voltages are treated as fixed inputs in the analysis, not just nodes 0 to $m-1$. Any one-port group within the element subnetworks that is not terminated on node i (considering only drain and source terminals for $\mathcal{N}$-elements and $\mathcal{P}$-elements) does not affect the ith node equation.

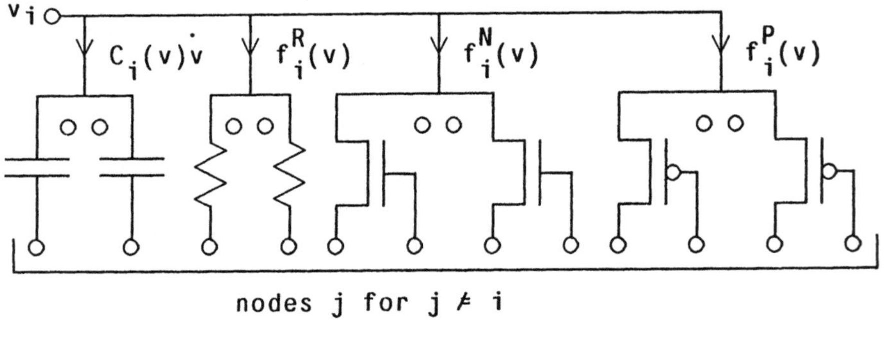

Figure 4-26: Circuit corresponding to single node equation.

Equation (4.6) is the node equation that corresponds to the circuit of figure 4-26 rewritten in normal form. On the right hand side of the equation the entire voltage vectors are used as arguments, but only the terms appear that correspond to nodes connected to node i through a one-port group of devices. The ith components of the functions g^C, f^R, f^N, and f^P do not depend on the rest of the voltages. In typical MOS VLSI circuits, internal nodes are generally directly connected to only a small number of other nodes and elements. The function g^C represents the incremental constitutive relation of $\mathcal{N}_C$ when only the ith terminal current is treated as an independent variable.

$$\dot{v}_i = g_i^C(v, [\dot{v}_0, .. \dot{v}_{i-1}, -f_i^R(v) - f_i^N(v) - f_i^P(v), \dot{v}_{i+1}, .. \dot{v}_{n-1}]^T). \qquad (4.6)$$

Based on the partial ordering of waveforms introduced in the last chapter, and the definitions of monotonicity for each subnetwork derived in the previous three sections, the solution of a node equation is monotonic in both its initial state and its inputs, including the characteristics of its associated devices. The main caveat is that when input nodes are connected to two different element subnetworks, their effect on each must be considered separately. As mentioned in section 3.2, such a monotonic property allows the behavior of a circuit such as the one pictured in figure 4-26 to be bounded by that of a simpler circuit. The result is derived in detail in the next two subsections, and an example is presented in the last one.

4.5.2 Differential Equation Inequalities

When solving a node equation over a time interval, a trajectory (or waveform) for the voltage or charge is determined. For exact analysis, a discrete approximation to the differential equation is often considered. When bounding the solution, though, the actual differential equation must be considered. A bound on an approximation does not generally provide a rigorous bound on the actual solution, and common numerical integration formulas do not lend themselves to simple and efficient bounding analysis. Analysis of the solution to the differential equation, though, can still be partitioned into small time intervals, producing similar time steps in the analysis. This subsection presents an inequality theorem for nonlinear differential equations that is used both in this section and in the next to derive monotonic properties for circuit behaviors.

Before the main inequality theorem is presented, note that the results from the last three sections can be used almost directly to produce simple bounds on the solution of equation (4.6). The only observation needed is that integration is a monotonic operator. Bounds on the inputs, elements, and v_i can be transformed into bounds on $\dot{v}_i$ by using the previous results. Bounds on v_i can then be integrated to bound v_i. There are two problems with this technique, though. First, initial bounds for v_i are required. As shown in the next chapter, initial bounds can either be easily guessed, or they are given during a relaxation procedure. Second, and more importantly, this method ignores correlations between v_i and $\dot{v}_i$ in the calculation of voltage trajectories. As a result, the bounds must always diverge. The restoring nature of the logic is discarded with the simplification gained by ignoring the correlations. This technique can still be useful in two different situations, though. First, as illustrated later in section 5.3, it can be used to initialize a bound relaxation algorithm. Second, it can be partially used where the correlations it ignores are not significant. For example, it is used in this section to treat the operating point calculations for the capacitor subnetwork in incremental analysis.

To derive more powerful bounding algorithms, a theorem proved in [52] and stated here as theorem 4-10 can be used. Roughly speaking, for a scalar equation the theorem says that if the scalar derivative function f in $\dot{x}=f(x,t)$ is nonnegative for $x=0$, trajectories for x(t) can never escape from the region $x \geq 0$. For the vector case, x(t) can never escape from the region $x \geq 0$ if each component $f_i(x,t)$ is nonnegative when its corresponding state component $x_i=0$ and the other state components are nonnegative. In other words, each state component is pushed back if it tries to leave the region first.

Theorem 4-10: Let $f: \mathbb{R}^n \times \mathbb{R}_+ \rightarrow \mathbb{R}^n$ be continuous. Let f(x,t) be globally Lipschitz in x uniformly[49] in t. If for each $i \in \{1,2,...n\}$ and all $t \geq 0$, $f_i(x,t) \geq 0$ $\forall x \geq 0$ with $x_i=0$, then for any solution[50] $x(\cdot)$ to $\dot{x} = f(x,t)$, $x(0) \geq 0 \Rightarrow x(t) \geq 0$ $\forall t \geq 0$.

To derive monotonic properties for the solutions of a differential equation, theorem 4-10 must be applied in an incremental sense. Consider the solutions x(t) and $\hat{x}(t)$ to the two equations $\dot{x}=f(x,t)$ and $\dot{\hat{x}}=\hat{f}(\hat{x},t)$ with the initial conditions $x(0)=x_0$ and $\hat{x}(0)=\hat{x}_0$. If the difference $\hat{x}(t)-x(t)$ is denoted $\delta(t)$, a differential equation to be solved for $\delta(t)$ can be derived by taking the difference of the two derivative functions.

$$\dot{\delta}=f(\hat{x}+\delta,t)-\hat{f}(\hat{x},t), \qquad y(0)=x_0-\hat{x}_0. \tag{4.7}$$

---

[49]Lipschitz in x uniformly in t means that the Lipschitz constant can be chosen such that it is not a function of t.

[50]The conditions imposed on the function f imply a unique solution for a given initial condition.

This difference function has the same continuity properties as each of the individual functions. If $\delta(t)$ is nonnegative, i.e., each component is nonnegative for all time, then the waveform $\hat{x}(t)$ is larger than $x(t)$. The following corollary is proved in [34] and restated here.

Corollary 4-11: Let $f,\hat{f}:\mathbb{R}^n \times \mathbb{R}_+ \rightarrow \mathbb{R}^n$ be continuous and globally Lipschitz in their first argument uniformly in their second. If for each $i \in \{1,2,...n\}$ and all $t \geq 0$, $\hat{f}_i(\hat{x},t) \geq f_i(x,t)\ \forall \hat{x} \geq x$ with $\hat{x}_i = x_i$, then for any solutions $\hat{x}(\cdot)$ to $\dot{\hat{x}} = \hat{f}(\hat{x},t)$ and $x(\cdot)$ to $\dot{x} = f(x,t)$, $\hat{x}(0) \geq x(0) \Rightarrow \hat{x}(t) \geq x(t)\ \forall t \geq 0$.

Based on the corollary, a partial ordering on derivative functions $f(x,t)$ can be defined. The corollary then says that the behavior of a differential equation as in (4.6) is a monotonic function of both its initial condition and its derivative function $f(x,t)$. The results of the previous three sections are used to derive conditions under which this derivative function is monotonic in both the inputs and elements of a circuit, resulting in similar monotonic properties for the solution.

4.5.3 Monotonicity of the Node Equation

To consider the implications of theorem 4-10 and its corollary to the analysis of the node equation, only the scalar case is needed. The previous result implies that a scalar behavior $x(t)$ for $0 \leq t \leq T$ is monotonic in both $x(0)$ and the derivative function $f(x,t)$, using a simple partial ordering on the derivative function. Therefore, to show monotonicity of the node equation, only monotonicity of the derivative function in

$$\dot{v}_i = g_i^C(v^A, [\dot{v}_0^B, .. \dot{v}_{i-1}^B, -f_i^R(v^C) - F_i^N(V^D) - F_i^P(V^E), \dot{v}_{i+1}^B, .. \dot{v}_{n-1}^B]^T), \tag{4.8}$$

must be shown, which is equivalent to equation (4.6) for exact analysis. The superscripts on the voltage vectors denote different instances of the same voltages as will be discussed momentarily. To discuss monotonicity, the extended voltage vectors are used for the transistor subnetworks.

The input v^A for the node equation is not considered directly in the analysis of this section. Bounds on the operating point for $\mathcal{N}_C$ are incorporated through the incremental capacitor values using only the simple technique discussed in the previous subsection[51]. As mentioned in the last section, it is assumed that v^A is treated as a constant over significant time intervals, so simple monotonic relationships between v^A and g_i^C are not required. Only simple relationships between the incremental capacitor values and g_i^C are needed. One way to look at this incorporation of v^A is that the circuit model is being relaxed. Initially, there is a wide range of incremental capacitor values possible for any time interval. However, as the bounds on the operating point shrink, so do the bounds on the incremental capacitors. By combining this technique with the relaxation algorithms discussed in chapter five, an additional relaxation of models can be intertwined with space and time relaxation. Model relaxation alone has been suggested in [35].

The results of the previous sections show that the derivative function is a composition of monotonic functions, and therefore monotonic itself with respect to the inputs and circuit elements. When voltage is used as a state variable, the derivative is calculated through a calculation of the constitutive relations of $\mathcal{N}_C$ with a particular input current at the node. As seen in section 4.4 the transfer function used is monotonic. The node input current is evaluated by adding the contributions of the resistive subnetworks. As seen in sections 4.2 and 4.3, these contributions are monotonic as well, with a slight generalization of the circuit in the case of the transistor subnetworks. The node voltage v_i has no instances in the extensions of the voltage vectors that affect F_i^N or F_i^P due to the lack of internal feedback paths in $\mathcal{N}_N$ and $\mathcal{N}_P$. The only new notion is that the node voltages (other than at node i) must be treated as different instances of the same voltages when they appear as arguments to more than one of the functions f_i^R, f_i^N, and f_i^P. This is just an extension of the generalization used inside the transistor subnetworks that is needed to include all resistive subnetworks together. The superscripts on the voltage vectors denote this independence. Only the component v_i is treated as the same quantity among all voltage vectors. As a result, the derivative function is "monotonic" in the input

---

[51] If feedback connections between a terminal of $\mathcal{N}_N$ and the gate of a one-port group connected to it were allowed, they could also be incorporated with this simple technique.

sources and in the devices of figure 4-26, and the solution of equation (4.6) can be bounded with the solution of a differential equation obtained by using extreme values for these input quantities. Of course, monotonic simplifications can even be used to reduce the equation to one with a closed-form solution.

There is another useful property of the derivative function for the node equation that can be used in bounding algorithms. It is easy to show that the derivative function (the right hand side of equation (4.9)) is a monotonically decreasing function of v_i. The three resistive subnetwork currents $f_i^R(\mathbf{v})$, $f_i^N(\mathbf{v})$, and $f_i^P(\mathbf{v})$ are monotonically increasing functions of v_i, so their sum is an increasing function of v_i. Due to the monotonicity of the terminal constraints of $\mathcal{N}_C$, the result follows.

A similar analysis can be done for a node equation using charge as a state variable. This time the derivative function is given by the resistive subnetworks evaluated for a particular node voltage. As before, the derivative function is monotonic with a reasonable definition of monotonicity. The node voltage is then obtained from a non-incremental analysis of the capacitor subnetwork, using the state variables as an input, along with other node voltages. In section 4.4 it is demonstrated that this mapping is also monotonic.

Theorem 4-12: Assume that the $v_j(t)$ are continuous for $t \in [0,T]$. The solution of equation (4.8) $v_i(t)$ for $0 \leq t \leq T$ is monotonic in the initial state $v_i(0) = v_{i0}$, monotonic in $\dot{v}_j^B$ for $j \neq i$, monotonic in v_j^C for $j \neq i$, monotonic in V_j^D for $j \neq i$, monotonic in V_j^E for $j \neq i$, current-wise monotonic[52] in each incremental one-port capacitor group, current-wise monotonic in each one-port resistor group, and current-wise monotonic in each one-port transistor group.

Proof: The function g_i^C is globally Lipschitz in its second argument by theorem 4-8. The derivative function is then Lipschitz in $\mathbf{v}^C$ by theorem 4-3 and in $\mathbf{V}^D$ and $\mathbf{V}^E$ by theorem 4-5. The node voltages are continuous in t as they are differentiable, and the node voltage

[52] Recall that special notions of monotonicity were used in sections 4.2 through 4.4.

derivatives are also continuous by assumption. Therefore, all continuity requirements of corollary 4-11 are satisfied. The function g_i^C is monotonic in its second argument by theorem 4-8, and current-wise monotonic in all one-port capacitor groups by corollary 4-9. The derivative function is then monotonic in v_j^C for $j \neq i$ and current-wise monotonic in all resistors by theorem 4-3. The third argument is also monotonic in V_j^D and V_j^E for $j \neq i$ and current-wise monotonic in all transistors by theorem 4-5. The result then follows from corollary 4-11. ∎

4.5.4 Analysis of the Node Equation

An example to present a typical application of the previous results is given in this subsection. Consider obtaining an upper bound on the solution of a circuit as pictured in figure 4-26 when voltage is used as a state variable. To do this, the solution of a different circuit must be obtained. An upper bound on the initial voltage is used along with a "bounding" circuit, i.e., a simplified circuit model with a behavior that bounds that of the original model.

The first step in deriving a bounding circuit is to consider the resistive subnetworks. The bounding circuit is derived using an upper bound for all other voltage waveforms v_j for $j \neq i$ except for instances connected to transistor gates. Upper bounds for these voltages could be piecewise continuous or even piecewise constant. Analysis of intervals can be connected through final and initial values of the state variables for each time interval. Upper bounds are used for the conductances of the resistors and transistors during times in which they are sending current into node i, and lower bounds are used at other times. A similar rule is used for the gate terminals of the transistors, depending on whether the transistors are n-channel or p-channel. Bounds on the devices can also be piecewise linear or piecewise constant[53].

The second step is to consider the capacitor subnetwork. The bounding circuit must use an upper bound on all of the input voltage derivative waveforms, which also can be simplified. Finally, if the capacitors are unknown, piecewise linear circuits can be evaluated to determine a bound. The

[53]Strictly speaking, the devices must be continuous, but solutions using continuous devices can generally approach arbitrarily close to a solution obtained using only piecewise continuous devices.

transformation mentioned in section 4.4 can be used to further simplify this computation. The transformations needed to bound a node circuit are roughly those that a circuit designer would expect based on intuition.

4.6 Cluster Circuits

The node equation presented in the previous section represents the most fine-grained partitioning that is possible in evaluation of the circuit node equations. Relaxation can be applied to such a partition to analyze an entire circuit. However, due to performance of the relaxation algorithm, a larger partition is often desired. The basic unit in a MOS circuit, when considering the coupling present between nodes, is the cluster circuit. Except when logic feedback paths exist, these clusters generally have strong bidirectional internal coupling and primarily one-way external coupling.

A cluster is defined as a group of nodes that is directly connected through a resistive path that passes only through cluster nodes. A resistive path is one in which d.c. current may flow. Figure 4-27 contains a typical cluster circuit that contains of a number of internal nodes. If an internal transistor is turned off during a particular time period, it is possible during that period to analyze the circuit as two smaller clusters.

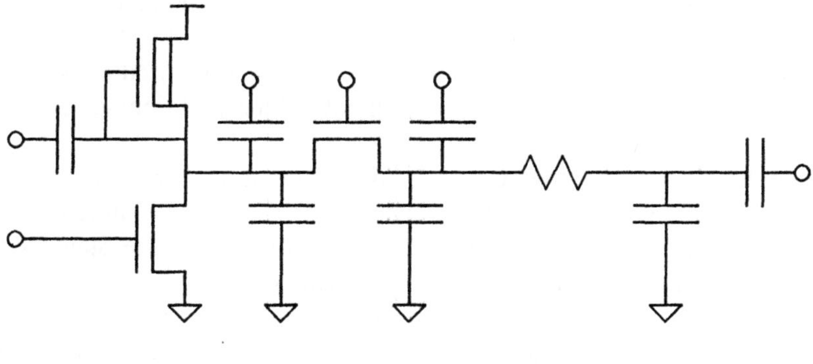

Figure 4-27: Circuit corresponding to a typical cluster.

Fortunately, the results for node subcircuits can be extended to general cluster circuits in most cases. If there is no capacitive coupling between internal nodes, and no internal feedback through transistor gates except when they are in "diode" configurations, as is often the case in cluster circuits, the entire cluster can be analyzed directly with monotonic relationships similar to those used for a node subcircuit. Only a simple circuit transformation is required which duplicates each resistive element that spans two internal nodes to separate its effect on each node. Only in the presence of either direct capacitive coupling between nodes or internal cluster feedback is partitioning below the level of clusters required.

There is one important special case of a cluster circuit that often appears in an intermediate-level model for which more powerful monotonic relationships exist. That is the model for a restoring logic gate, NMOS or CMOS, containing no Miller capacitance, no capacitance within the pullup or pulldown networks, and an RC tree model for interconnect load. Such a model is also considered in this section.

Linear RC circuits represent a very special cluster circuit used in simple MOS circuit models for which a considerable amount of bounding analysis has been done. The bounding work mentioned in the second chapter represents treatment of even simpler special cases of cluster circuits. The work on one-way gate models presented here is a direct extension of that work.

4.6.1 General Cluster Circuit

When considering a cluster of nodes simultaneously, monotonic properties of the solution can be derived by appealing to the complete vector version of theorem 4-10 and its corollary. A vast majority of cluster circuits found in practice can be considered with the results derived in this subsection. For those that violate the restrictions placed on the circuit, a finer partitioning can always be used.

In the vector case of theorem 4-10, the solution increases if the initial condition increases and each component of the derivative function both increases and is monotonic in the other components of the state vector. Roughly speaking, none of the components can fall below their old values because each is trapped above as long as the others remain above. When any one component or group of components approaches its old value, the conditions of theorem 4-10 push them away.

Consider the set of differential equations that describe a cluster containing r nodes labeled i for $m \leq k \leq i \leq k+r-1 \leq n-1$:

$$\dot{v}_i = g_i^C(\,v^{A_i}, [\dot{v}_0^{B_i}, \ldots \dot{v}_{i-1}^{B_i}, -f_i^R(v^{C_i}) - F_i^N(V^{D_i}) - F_i^P(V^{E_i}),$$
$$\dot{v}_{i+1}^{B_i}, \ldots \dot{v}_{n-1}^{B_i}]^T\,). \tag{4.9}$$

As before, superscripts represent different instances of the input voltage vectors. Only the components v_i for $k \leq i \leq k+r-1$ are necessarily identical in all vectors. In the last section it was shown that each equation in (4.9) has a right hand side that is monotonic in its inputs and in its elements when considered alone, but now the interaction of the equations must also be considered.

As with the single node equation, the effect of input voltages on the different element subnetworks is treated separately. For cluster circuits, the effects on more than one node equation must also be considered. For an input that is not connected to cluster nodes through gates of transistors, each component of the derivative function is a monotonically increasing function of its voltage. Creating separate instances of each does not create an ignored correlation. Various gate voltages can have different effects on each component, but they are already separated both within and among the transistor subnetworks. A problem arises only when a single instance of a gate voltage has opposite effects on two components, and this can only happen when it is connected to a transistor that spans the two corresponding nodes. Any resistive element that spans two internal cluster nodes has the same problem with respect to monotonicity in the element. The solution is to transform any such element as pictured in figure 4-28 for the case of a resistor. Each instance of the element is treated separately. In the case of a transistor, the gate voltage source is also duplicated and the instances are treated separately[54].

The next concern is the effects of capacitive coupling inside a cluster circuit. Each component of the derivative function is a monotonically increasing function of the input voltage derivative sources associated with input nodes that are directly capacitively coupled to internal cluster nodes. If two internal cluster nodes are coupled to each other, though, there is no way to guarantee the effect on the component associated with one node over wide ranges of voltages for the other nodes. As a result, internal capacitive coupling, which rarely exists in practice, is not allowed. In terms of equation (4.9), this means that no component $\dot{v}_i$ is a function of any input $\dot{v}_j$ for $k \leq j \leq k+r-1$.

The final concern is the effect of other cluster node voltages on each

---

[54]Since this transformation can only be made for lumped element models, distributed capacitance cannot be considered without further results.

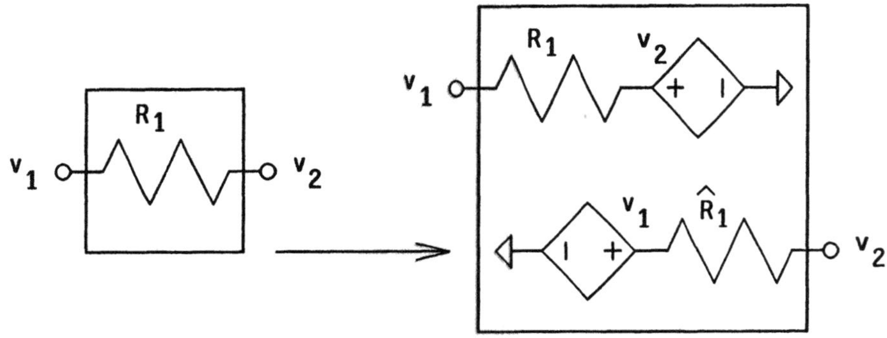

Figure 4-28: Resistive elements between cluster nodes can be transformed.

component of the derivative function. When coupling is only through resistive elements that span two internal nodes, the relationship is always monotonically increasing as required by theorem 4-10. A problem arises only when one of the transistor gates associated with one cluster node is connected to another cluster node. Such rare feedback, an example of which can be found in the Schmitt trigger in figure 3-16, makes the relationship uncertain and is not allowed in the cluster circuits considered here.

> **Theorem 4-13:** Let L be the set of all integers i such that $k \le i \le k+r-1$. The solution of a general cluster circuit as defined in this subsection (described by equation (4.9)), $v_i(t)$ for all $i \in L$ and $\forall t \in [0,T]$, is monotonic in the initial state ($v_i(0)=v_{i0}$ for all $i \in L$), is monotonic in $\dot{v}_j^B$ for $j \notin L$, is monotonic in v_j^C for $j \notin L$, is monotonic in V_j^D for $j \notin L$, is monotonic in V_j^E for $j \notin L$, is current-wise monotonic[55] in each incremental one-port capacitor group, is current-wise monotonic in each one-port resistor group, and is current-wise monotonic in each one-port transistor group.

---

[55] Recall that special notions of monotonicity were used in sections 4.2 through 4.4.

Proof: To extend the proof of theorem 4-12 to consider multiple node equations simultaneously, the full vector results of corollary 4-11 must be used. The continuity arguments are similar. Monotonicity in the inputs and in the elements follows by considering each node equation separately. None share elements or instances of inputs as a result of the assumptions. To satisfy the requirements of corollary 4-11, it must also be shown that the right hand side of each node equation is a monotonically increasing function of the voltages of all other nodes inside the cluster. Pick any node $i \in L$ and another node $j \in L$ such that $j \neq i$. f_i^R is a monotonically decreasing function of v_j as a result of theorem 4-3, f_i^N and f_i^P are monotonically decreasing functions of v_j as a result of theorem 4-5 and the restriction against internal feedback paths in the cluster. Since the function g_i^C is a monotonically increasing function of its second argument, as a result of theorem 4-8, the required monotonicity follows. ∎

The transformation of figure 4-28 is a generalization of the technique used in [35] to extend bounding results for linear RC trees to include nonlinear resistors. This interpretation of the algebra used in that result provides additional intuitive understanding of the technique. When the method is extended to treat transistors, a significant amount of uncertainty amplification can occur. If there is a large amount of uncertainty in the state of a transistor ("on" or "off"), the correlation between its effect on each terminal node is important. It can even create situations such as one where the transition of a complex logic gate is bounded by that of an inverter created by shorting out all but one of its transistors. This produces a bound that is not as tight as is often desired. If the cluster is partitioned further, the transform is done implicitly by the relaxation algorithm when it ignores the correlations among blocks, which in the case of a partitioned cluster circuit can be significant. A slightly more sophisticated approach for treating clusters is needed to achieve very tight bounds in the general case. The current being lost in the transformed elements must be considered.

More powerful direct monotonic properties that avoid the transformation of figure 4-28 are only possible for special classes of cluster circuits. An important special case is considered in the next subsection. Direct monotonic relationships between internal cluster elements and cluster behavior do not generally hold.

4.6.2 One-Way Restoring Logic Gate

One special case of a cluster circuit that is very important for digital MOS circuits is the one-way restoring logic gate model. The most general model of a logic gate contains the problems of a general cluster in terms of producing monotonic relationships. There are very common intermediate range models, however, that do not have these problems. Roughly speaking, these models do not contain any internodal coupling capacitance, nor do they contain any grounded capacitance within the pullup and pulldown networks. These effects are modeled with a load capacitor. The pullup and pulldown networks must be series-parallel transistor groups, but can contain very sophisticated d.c. transistor models. The interconnect and next-stage load is modeled by a standard RC tree circuit. Circuits constructed exclusively from such models can be efficiently bounded without relaxation, since coupling between clusters is exclusively one-way.

4.6.2.1 Resistive Logic Gate Models

The pullup and pulldown networks in MOS logic gates can be modeled by the general *P*-elements and *N*-elements of section 4.3, but the entire gate cannot. A new *G*-element must be defined to model the logic gate that consists of the union of pullup and pulldown transistor networks. A *G*-element that contains only a single pullup transistor and a single pulldown transistor, i.e., an inverter, is referred to as a simple *G*-element. Voltages are restricted to the range of GND to VDD, assumed to be 0 and 5 volts with no loss of generality, as the one-way gate model cannot produce node voltages beyond this range[56].

The partial ordering of *G*-elements used considers the output current, i_{OUT}, as a function of all variable terminal voltages. For convenience, a partial ordering is also defined for outputs of *G*-elements alone that considers the output current as a time-varying function of output voltage only. Given fixed input voltage waveforms, the two orderings are consistent, i.e., a larger *G*-element has a larger output.

---

[56]This property can be verified by appealing to corollary 4-11. The actual solution can be compared to the solution obtained when all voltages are fixed at one extreme over the entire time interval.

Definition 4-14: A <u>*G*-element</u> is a multi-terminal resistor constructed from two general *N*-elements or from a general *N*-element and a general *P*-element as pictured in figure 4-29. Substrate terminals are assumed to be tied to the appropriate extreme input voltages, and the gates of both groups are labeled 1 to m. A <u>simple *G*-element</u> is a *G*-element constructed from only two transistors, one pullup and one pulldown. Given two (m+2)-terminal *G*-elements G_1 and G_2 characterized by $i_{OUT}=G_1(v_{G1}, .. v_{Gm}, v_{OUT})$ and $i_{OUT}=G_2(v_{G1}, .. v_{Gm}, v_{OUT})$, we say that $G_1 \geq G_2$ iff $G_1(v_{G1}, \ldots v_{Gm}, v_{OUT}) \geq G_2(v_{G1}, \ldots v_{Gm}, v_{OUT}) \; \forall \; v_{G1}, \ldots v_{Gm}, v_{OUT} \in [0,5]$. Given two time-varying *G*-element outputs O_1 and O_2 characterized by $i_{OUT}=H_1(v_{OUT},t)$ and $i_{OUT}=H_2(v_{OUT},t)$ respectively, we say that $O_1 \geq O_2$ iff $H_1(v_{OUT},t) \geq H_2(v_{OUT},t)$ $\forall v_{OUT} \in [0,5]$, and $\forall t \in [0,T]$.

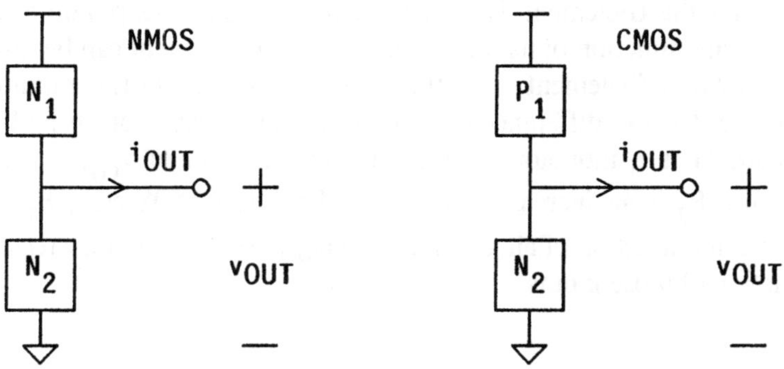

Figure 4-29: The transistor portions of MOS logic gates are *G*-elements.

While a *G*-element is not a general *N*-element or *P*-element, it maintains similar monotonicity properties. First, a *G*-element is a monotonic function of its pullup and pulldown subnetworks. Second, its current function is monotonic in all its terminal voltages. As a result of the second property, an output of a *G*-element produced by bounds on its inputs constitutes a bound on its output. These properties are listed in the following theorem.

Theorem 4-14: A *G*-element is an $(m+2)$-terminal resistor characterized by $i_{OUT}=G(v_{G1},..v_{Gm},v_{OUT})$, where G is continuous and is a monotonically decreasing function of all the inputs if all port voltages are within the range of 0 to 5 volts. Furthermore, a *G*-element is a monotonically increasing function of its pullup network and a monotonically decreasing function of its pulldown network.

Proof: Assume first that a *G*-element is constructed from two *N*-elements as in figure 4-27. By applying KCL at the output node, it follows that $i_{OUT}=F_1(5,v_{OUT}) - F_2(v_{G1},..v_{Gm},v_{OUT},0) = G(v_{G1}, .. v_{Gm},v_{OUT})$. $0 \leq v_{OUT} \leq 5$ so both F_1 and F_2 must have non-negative values $(v_D \geq v_S)$. Therefore, $F_2(v_{G1}, .. v_{Gm},v_{OUT},0)$ is a monotonically increasing function of N_2 and $F_1(5,v_{OUT})$ is a monotonically increasing function of N_1. As a result, the function G, and thus the *G*-element, is an increasing function of its pullup and a decreasing function of its pulldown. A similar argument can be made for a CMOS *G*-element. In either case, the function G is continuous because it is the difference of continuous functions. For an NMOS circuit, G is a monotonically decreasing function of $v_{G1}, .. v_{Gm}$ because F_2 is an increasing function. For the CMOS case, F_1 is a decreasing function of all gate input voltages so G is also a decreasing function of those inputs. ∎

In summary, the output of a *G*-element is a monotonically decreasing function of all the voltage waveforms at its gate terminals. This property is a reflection of the basic inverting nature of MOS logic gates. In addition, it is a monotonically decreasing function of the transistors in its pulldown network and a monotonically increasing function of the transistors in its pullup network.

4.6.2.2 Interconnect Models

R-elements and *C*-elements can be used to construct models for gate interconnections that include both the wires and the input loads of other gates.

The connection of *R*- and *C*-elements into an RC tree [9], a very general interconnect model, is represented as an I-element. Roughly stated, an RC tree is a tree structure of resistances, emanating from an input node and containing a grounded capacitor at each node. Any number of nodes, including the input node, can be considered output terminals. Usually the capacitors model either logic gate inputs or wire capacitance while the resistors model wire resistance. In circuits where wire lengths are very small, the interconnect model is often reduced to a single capacitor, a special case of an RC tree model.

Definition 4-15: An <u>I-element</u> is an RC tree containing *R*-elements and *C*-elements as pictured in figure 4-30. The *R*-elements must form a tree structure in which every internal node has a unique path through *R*-elements to a single input node. Every internal node contains a *C*-element between itself and the ground node, and can be considered an output node of the I-element.

I-elements also exhibit some important monotonic properties. If an I-element is driven by a gate output at its input, all output voltages are monotonic functions of the gate output and the initial state of the I-element. If the I-element is driven by a gate output that is a monotonic function of time, and the initial state produces monotonic outputs, another important property holds. All outputs, in addition to being monotonic functions of time themselves, are then monotonic functions of the *C*-elements. The theorem that expresses these properties is a simple extension of some existing theorems on the behavior of RC circuits [34].

Definition 4-16: Given two initial states $S=\{v_1(0), .. v_k(0)\}$ and $\hat{S}=\{\hat{v}_1(0), .. \hat{v}_k(0)\}$, we say that $S \geq \hat{S}$ iff $v_j(0) \geq \hat{v}_j(0) \forall j=1, .. k$. An initial state *S* of an I-element is said to be <u>consistent with a gate output *O*</u> iff the response of the I-element driven by *O* and with initial state *S* produces a state which is a monotonic function of time. In other words, the voltage waveforms at all internal nodes of the I-element must be all rising waveforms or all falling waveforms.

Theorem 4-15: All outputs of an I-element driven by a continuous

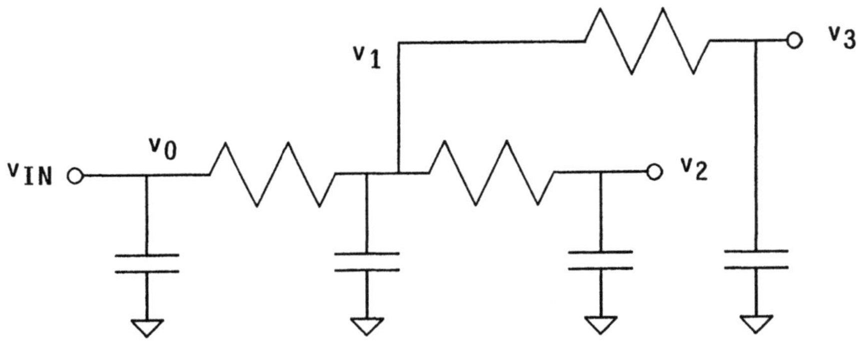

Figure 4-30: Gate interconnect wires are modeled with an RC tree.

gate output are monotonic functions of the gate output and the initial state. Furthermore, if an I-element is driven by a continuous gate output that is an increasing function of time, all I-element output voltages are decreasing functions of each internal *C*-element when the initial state is consistent with the gate output. Similarly, if the gate output is a decreasing function of time, all outputs are increasing functions of each *C*-element.

Proof: In [34] it is shown that this result is true for RC lines driven by a voltage source by using theorem 4-10. This result can be generalized to RC trees as mentioned in [34] and proven in [53], [54], and [55]. The result can also be easily generalized to consider a tree driven by a *G*-element output. The current term contributed by the source becomes $F(f_j(q_j),t)$ and all conditions for application of theorems still hold. ∎

4.6.2.3 One-Way Combinational Logic Models

G-elements and I-elements can be used to construct a model for an entire combinational logic circuit. A single *G*-element / I-element pair, where the output of the *G*-element is connected to the input of the I-element, is called a simple *L*-element. A general *L*-element, or logic circuit, is constructed from simple *L*-elements where each input of simple *L*-elements is connected to either a single output of another simple *L*-element or is considered an input to the entire *L*-element. Feedback paths are not allowed in *L*-elements.

Definition 4-17: A simple *L*-element consists of a *G*-element and an I-element with the output of the *G*-element driving the input of the I-element. An *L*-element is an acyclic directed graph where each node is a simple *L*-element as pictured in figure 4-31. The nodes have multiple inputs and multiple outputs, and no more than one arc terminates on each input. Any number of arcs may begin on a single output. Inputs with no arcs are considered to be inputs of the entire *L*-element and every output of each simple *L*-element is considered to be an output of the entire *L*-element.

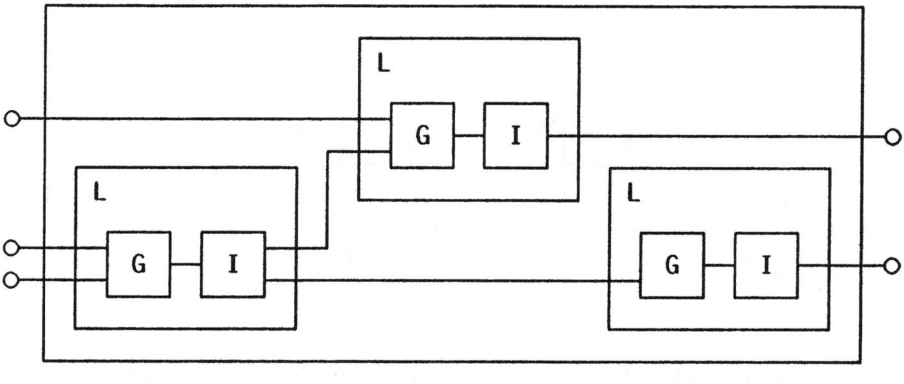

Figure 4-31: Each logic gate is modeled as a simple *L*-element.

The monotonic properties of a simple *L*-element are stated here as theorem 4-16. These basic results follow immediately from the previous theorems. The use of this theorem to generate bounds on simple *L*-elements is considered in chapter six.

Theorem 4-16: The output waveforms of a simple *L*-element are monotonic functions of the input waveforms, the internal transistors, and the initial state of the internal capacitors. In the case where the gate output is monotonic in time and the initial state is consistent, the output waveforms are also monotonic functions of the internal capacitors.

Proof: The first two relations follow from theorems 4-14 and 4-15. The last two follow directly from theorem 4-15. ∎

Since the outputs of each simple *L*-element are monotonic functions of the inputs, the responses of the element to bounds on its inputs constitute bounds on its outputs. As a result, the logic gates in the model are able to propagate bounding information, analogous to the way they propagate logic values. A bound on a general *L*-element can be obtained by composing those on individual simple *L*-elements.

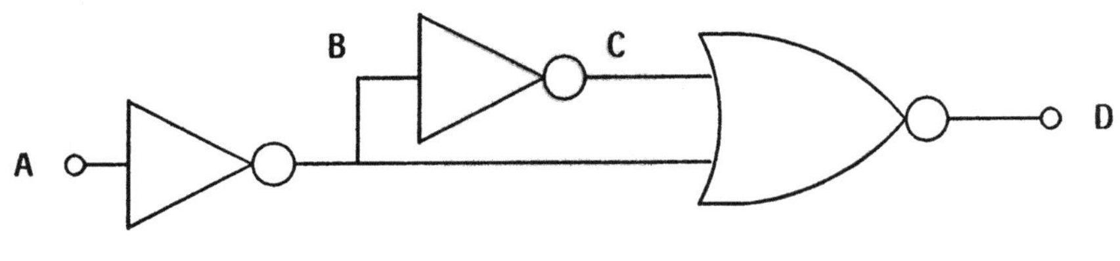

Figure 4-32: Effect of reconvergent paths in an *L*-element.

There are two reasons why the analysis of a general logic circuit is considered at the level of its composite simple *L*-elements. First, simple *L*-elements are the basic building blocks for arbitrary, unstructured logic circuits. Therefore, they provide the smallest degree of refinement that can be applied in general. Simple *L*-elements themselves are constructed from a very structured combination of more refined circuit elements. Second, complex *L*-elements do not have a monotonic dependence of outputs on inputs in general because they may contain reconvergent paths. Consider the circuit pictured in figure 4-32.

The response of an upper bound on the voltage waveform at node A produces a lower bound at node B, and an upper bound at node C. To generate an upper bound on the voltage waveform at node D, lower bounds are needed at both nodes B and C. Still, there are many cases where complex L-elements can be considered as a single element that produces a bound on its output in response to a bound on its input. A simple chain of inverters certainly has an output that is a monotonic function of its input because they are related by a simple composition of monotonic functions. Such a property can be useful in generating bounding macromodels for portions of a circuit.

4.7 Summary

The mathematical lumped element network models often used for MOS VLSI circuits contain a wealth of monotonic relationships that can be exploited to generate bounds on their behavior. Only very mild constraints need be placed on the element models that are satisfied by most reasonable models used in practice. The composition rules required to guarantee the useful monotonic relationships maintain a large degree of generality. Ironically, the most complex circuit models generally satisfy the composition rules by having at least a grounded capacitor and some resistive element connected to each internal node. The composition rules are only of concern in the realm of intermediate level models, to indicate where complexity can be relaxed without eliminating the ability to generate efficient bounds.

By considering the different device models separately, an algorithm for bounding the behavior of the network at each point in time can be divided into smaller problems. In addition, by adding some mild constraints to the network model, relaxation can be used to further partition evaluation of the network to a number of small subcircuits with only local connections. Waveform relaxation techniques allow this partitioning to include entire time intervals.

A very desirable subcircuit for partitioned analysis, from the point of view of the convergence rate of a relaxation algorithm, is a cluster subcircuit. These subcircuits, consisting roughly of internal nodes connected directly through resistive paths, often have strong bidirectional internal coupling. The results presented in this chapter can be used to bound the response of general cluster circuits, as well as important special cases. If only unidirectional coupling exists between cluster subcircuits, partitioned analysis can be accomplished simply by propagating bounds from inputs to outputs of clusters. In general, though, coupling is bidirectional and the relaxation techniques presented in the next chapter must be considered for partitioned analysis.

5. BOUND RELAXATION

Exact waveform relaxation algorithms for VLSI circuit simulation can be extended to treat waveform intervals. The first section of this chapter discusses the theory of such an extension, and the remaining sections cover its applications to circuit simulation. Any convergent iterative algorithm has an analogous interval algorithm that can be used both to generate and tighten bounds on a solution. Based on this observation, relaxation techniques can be used to generate efficient bounds on the d.c. and transient solutions of complex digital MOS circuits. A bounding algorithm for the network solution can thus be partitioned into a series of bounding algorithms for simple subcircuits, such as those considered in chapter four.

One part of the VLSI circuit simulation task is d.c. analysis, which is often used to supply the initial conditions for transient analysis. Since this often constitutes a fairly small portion of the full simulation time, bounding algorithms are not urgently needed to simplify d.c. computation. Bounding algorithms can, however, allow the incorporation of uncertainty. The most interesting aspect of d.c. analysis is that digital circuits often do not have a unique solution. Additional information, e.g., the desired initial states for latches, must be provided. Bounding algorithms can also be used to discover occurrences of multiple solutions. Even though conventional exact relaxation algorithms do not generally converge in d.c. analysis, relaxation methods can still be used to tighten bounds on a d.c. solution. Although relaxation methods cannot be used to generate initial bounds on a d.c. solution, they are not needed because the power supply voltages generally supply a valid d.c. bound. Bound relaxation algorithms can be used to tighten this rough initial bound to one of arbitrary accuracy in many cases. The second section of this chapter discusses the use of bounds in d.c. analysis.

The most important part of the VLSI circuit simulation task is transient analysis, as it constitutes a large portion of the full simulation time. For large and complex circuit models, bound relaxation algorithms are potentially very useful. The Waveform Relaxation algorithm for exact transient analysis can be extended to the interval case. An initial conservative guess for the solution can be tested for validity as a bound by the extended algorithm, as well as tightened by relaxation iterations. As argued in chapter three, the performance of such an algorithm is similar in many cases to the analogous exact procedure. The result can be a telescoping sequence of bounds that can be calculated until a desired level of accuracy, within some range, is obtained. The third section of this chapter discusses the use of bounds in transient analysis.

A common simple model for MOS circuits is a linear RC mesh network. Closed-form bounds have been derived for certain linear RC networks and have been used for the analysis of these models. General relaxation techniques can also be used on these circuits since they are special cases of the more general model. Due to the special properties of these models, though, more powerful results can be shown. For a very general class of linear RC circuits, with arbitrary initial conditions, a relaxation algorithm will produce a sequence of closed-form bounds that achieves arbitrary accuracy. The last section of this chapter discusses the application of bound relaxation to the analysis of linear RC mesh circuits.

5.1 Theory

The theory behind the combination of relaxation and bounding is presented here in a slightly more general context. Instead of restricting the analysis to relaxation functions, more general iteration functions are considered. These iteration functions can operate on solutions over entire time intervals in the most general case. The definition for circuit behavior used in chapter three is generalized here to include any set of variables used by the iteration function. The old definition is appropriate in the special case where the Waveform Relaxation algorithm provides the iteration function and voltage is used as the state variable.

5.1.1 Definitions

A circuit solution, approximate or exact, consists of a set of responses for various circuit variables. All circuit variables of interest are considered to be output components of a vector-valued function, called a circuit behavior, defined on a given time interval. The circuit solution consists of a unique element, w^*, in the set of all possible circuit behaviors, $\mathcal{W}$.

Definition 5-1: A circuit behavior w is a continuous function w:$[0,T] \rightarrow \mathbb{R}^n$ (and therefore bounded) where each component[57] $w_i(t)$ for $0 \leq t \leq T$ represents a circuit variable in some network for $0 \leq t \leq T$. The set of all such circuit behaviors is called $\mathcal{W}$.

The iteration function, denoted F, operates on the vector-valued circuit behaviors in $\mathcal{W}$. The iteration function must have the property that given any initial circuit behavior w^0, the sequence of behaviors $\{w^0, F(w^0), F(F(w^0)), \ldots\}$ converges to a unique circuit behavior that represents the circuit solution. The circuit solution, w^*, must also be a "fixed point" of the iteration function, i.e., $F(w^*) = w^*$.

Definition 5-2: An iteration function is a function F:$\mathcal{W} \rightarrow \mathcal{W}$ where for any $w^0 \in \mathcal{W}$, the sequence $\{w^0, w^1, w^2, \ldots\} = \{w^0, F(w^0), F(F(w^0)), \ldots\}$ $= \{F^{(0)}(w^0), F^{(1)}(w^0), F^{(2)}(w^0), \ldots\}$ converges uniformly[58] to a unique element $w^* \in \mathcal{W}$, independent of w^0, such that $w^* = F(w^*)$. The notation $F^{(m)}$ denotes the composition of F with itself m times.

The Waveform Relaxation mapping satisfies the previous definition of an iteration function. In this case the iteration function consists of a series of solutions for all network subcircuits. The given initial state of the circuit is used

---

[57] A subscript after a circuit behavior denotes a component while a superscript denotes an index in a sequence of circuit behaviors.

[58] For the results pertaining to circuit simulation presented here, pointwise convergence is sufficient. However, Waveform Relaxation guarantees the stronger notion of uniform convergence and this suggests a more natural abstract generalization presented later.

in all calculations. Such a mapping for each circuit variable meets the requirements[59] of an iteration function F for the circuit model presented in chapter four.

A bound on a circuit behavior in the set $\mathcal{W}$ is specified with two other elements of $\mathcal{W}$ that represent the minimum and maximum values of its components during the time of analysis. These lower and upper bounds, named L and U respectively, define a closed circuit behavior interval $W = [L, U]$. A bound on a circuit behavior consists of a circuit behavior interval, W, within which it must fall. The interval definition used here is a simple extension of the one used in chapter three.

Definition 5-3: A circuit behavior $w \in \mathcal{W}$ is said to be <u>greater than or equal to</u> $\hat{w} \in \mathcal{W}$ iff $w_i(t) \geq \hat{w}_i(t)$ $\forall t$ such that $0 \leq t \leq T$ and $\forall i$ such that $1 \leq i \leq n$. A <u>behavior interval</u> $W = [L, U]$, $L \leq U$, is the closed set of circuit behaviors $\{w \in \mathcal{W} : L \leq w \leq U\}$ where $L \in \mathcal{W}$ and $U \in \mathcal{W}$.

The iteration function F represents a "point" map on the set of all circuit behaviors. In this set, each "point" actually corresponds to an entire circuit behavior. A bounding function, or bound on an iteration function, is expressed with an "interval map." All intervals are of the form just defined and the set of all such intervals is called I. An interval map is a map from I to I. A bounding function, denoted G, is an interval map that is defined relative to a specific iteration function. Given a bound on the input to the iteration function in the form of a behavior interval, the bounding function produces a bound on the output. The circuit behavior interval $G(W)$ is guaranteed to contain the image of interval W under the function F. The image of W under the function F is not required to be an interval, based on the assumptions given for the function F, but it is a subset of the interval $G(W)$. Figure 5-1 contains a simplified illustration with circuit behaviors represented as individual points on a vertical line. Note that $G(W)$ can contain additional behaviors that are outside the image of W.

---

[59]The Waveform Relaxation convergence proof assumes that the initial guess w^0 is consistent with the initial state of the circuit. However, by the definition [11] of the iteration function, w^1 will be consistent even if w^0 is not, so the sequence must still converge.

Definition 5-4: A <u>bounding function</u> G for an iteration function F is a mapping G:I→I such that ∀ $w \in W \in$ I, $F(w) \in G(W)$.

Many bounding functions have the property that shrinking their input interval will always shrink their output interval or leave it unchanged. A bounding function with this property is said to be inclusion monotonic [56]. Also, two bounding functions can often be compared by the tightness of the bounds they produce. One bounding function is said to be tighter than another if it produces a tighter bound for all possible inputs.

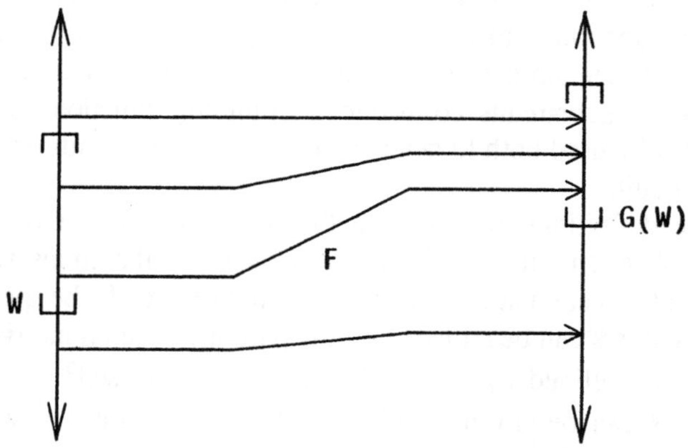

Figure 5-1: For any interval W, F maps any element in W to one in $G(W)$.

Definition 5-5: An <u>inclusion monotone bounding function</u> has the property that for all circuit behavior intervals W and $\hat{W}$ such that $W \subseteq \hat{W}$, $G(W) \subseteq G(\hat{W})$. G is said to be <u>tighter than or equal to</u> $\hat{G}$ if, for all circuit behavior intervals W, $G(W) \subseteq \hat{G}(W)$.

A bounding function based on the Waveform Relaxation algorithm bounds the behavior of each subcircuit separately. For a Gauss-Jacobi relaxation algorithm, $G([L, U])$ consists of separate bounds on the variables in each subcircuit, as calculated by assuming all other variables fall within the range given by $[L, U]$. The calculations must also assume that the initial state of the circuit lies within some given bound. Direct bounds for the input-output behavior of small subcircuits can be used to calculate a bounding function for the iteration function used in Waveform Relaxation. Given a set[60] of easily calculated bounding functions (G) for an iteration function (F), the practical goal is to find a sufficiently small circuit behavior interval that is guaranteed to contain the circuit solution w^*.

The most important properties of bounding functions are explored in the next two subsections. The first shows that repeated application of a bounding function to a bound on the solution can tighten the bound. The second shows that bounding functions can be used to generate bounds by testing if circuit behavior intervals contain their images under an iteration function. Since direct bounds on VLSI subcircuits constitute a bounding function for the entire circuit, they can be used both to generate and to tighten bounds on the solution of the entire circuit.

Though the definitions presented in this subsection are tailored to circuit simulation applications, the results in the next two subsections remain valid when the definitions are made more general and abstract. To be specific, the set of circuit behaviors $\mathcal{W}$ can be replaced by any partially ordered metric space such that the intervals defined by the partial ordering are closed[61]. The iteration function $F:\mathcal{W}\rightarrow\mathcal{W}$ can be any map with a unique fixed point x^* such that the sequence of images $\{F^{(m)}(x)$ for $m=1,2,\ldots\}$ of any point $x\in\mathcal{W}$ converges to x^*. As before, the bounding function G can be defined as a map on intervals in the partially ordered metric space. While the results in the next two subsections are presented in the circuit simulation context, they are valid in this more general context as well.

---

[60] In practice there are generally many different bounding functions that can be derived for an iteration function, ranging from tight and complex to loose and simple.

[61] The partial ordering on the metric space $\mathcal{W}$ is only used to express bounds in the convenient form of a closed interval, instead of a general closed set.

5.1.2 Using Bounding Functions to Tighten Bounds

The first property of a bounding function is used in an algorithm to tighten an existing bound on a circuit solution. It is based only on the circuit solution being a "fixed point" of the iteration function. Assume that a circuit behavior interval is given that is known to contain the circuit solution w^*. In other words, a bound on the circuit solution is known. If a bounding function is applied to this initial circuit behavior interval, the circuit solution is guaranteed to lie in the resulting output interval. Since the iteration function maps the circuit solution to itself, this follows from the definition of a bounding function. As a result, given an initial bound on the circuit solution W^0, $W^0 \cap G(W^0)$ produces a new bound that is tighter than or equal to the original. By repeating this process, a sequence of bounds can be produced that never expands. In fact, if $G(W^0) \subseteq W^0$ and G is inclusion monotone, the telescoping, i.e., never expanding, sequence is simply $\{W^0, G(W^0), G(G(W^0)), \ldots\}$. This first property was demonstrated in section 3.2 for a special case of bound relaxation, and is proven here for the general case.

Theorem 5-1: Assume $w^* \in W^0$ and let $W^{i+1} = W^i \cap G(W^i)\ \forall\ i \geq 0$. Then $w^* \in W^{i+1} \subseteq W^i\ \forall\ i \geq 0$. Furthermore, if G is inclusion monotone and $G(W^0) \subseteq W^0$, then $G(W^i) \subseteq W^i\ \forall\ i \geq 0$.

Proof: Let $w^* \in W^0$ and let $W^{i+1} = W^i \cap G(W^i)\ \forall\ i \geq 0$. For an inductive argument, assume $w^* \in W^k$ for some $k \geq 0$. Then $F(w^*) = w^*$ is an element of $G(W^k)$ by the definition of the function G. Therefore, $w^* \in W^k \cap G(W^k) = W^{k+1} \subseteq W^k$. Then by induction, $w^* \in W^{i+1} \subseteq W^i\ \forall\ i \geq 0$. Now, in addition, let G be inclusion monotone and $G(W^0) \subseteq W^0$. For an inductive argument, assume $G(W^k) \subseteq W^k$ for some $k \geq 0$. Then $W^{k+1} = W^k \cap G(W^k) \subseteq W^k$. Therefore, $G(W^{k+1}) \subseteq G(W^k)$ because G is inclusion monotone. Then by induction, $G(W^i) \subseteq W^i\ \forall\ i \geq 0$. ∎

If a loose bound on the circuit solution is available, it must be tightened to the degree required in a particular application. As just illustrated, any bounding function can be used to tighten bounds on the solution through iteration. In particular, inclusion monotone bounding functions can, under the right conditions, directly produce a sequence of telescoping bounds. For a particular bounding function, there is a limit to the degree of accuracy that can be obtained. If more accuracy is required, a tighter bounding function must be used. Both the number of iterations undertaken and the bounding function used provide the means to trade speed for accuracy. In general, a tighter bounding function will require more computation. If a bounding function is based on the Waveform Relaxation iteration function, new bounds need not be computed for all subcircuits in each iteration of a tightening algorithm. Since the iteration function is partitioned by subcircuits, new tighter bounds can be calculated only for certain variables over time intervals where their bounds are too loose. Using the old bounds on other variables is guaranteed to be conservative since the algorithm never produces looser bounds.

In practice, if an iteration function converges quickly to a solution, a tight bounding function will converge at a similar rate towards its optimal bound. The simple examples in section 3.2 demonstrate this property in a special case. In general a tight bounding function produces a good approximation to the image of its input interval, so any contractive properties of the iteration function are maintained in the bounding function. Since the Waveform Relaxation function is usually highly contractive for digital MOS circuits, a tight bounding function should, for the most part, shrink a loose bound quickly towards its optimal bound. In cases where a relaxation function is weakly contractive, a loose bounding function can lose the contractive property in some domain, and thus the ability to tighten a loose bound in all cases.

5.1.3 Using Bounding Functions to Generate Bounds

The second property of a bounding function is used in an algorithm to generate a bound on a circuit solution. It is based on the concept of "introverted" sets - those that are guaranteed to contain their image under an iteration function, F, or some number of compositions of F. An introverted circuit behavior interval is guaranteed to be a rigorous bound on the circuit solution.

> **Definition 5-6:** A circuit behavior interval W is said to be <u>introverted under the iteration function F</u> if there exists an integer m>0 such that $F^{(m)}(W) \subseteq W$. As in definition 5-2, $F^{(m)}$ denotes the composition of F with itself m times.

Assume that a bounding function is iterated as illustrated in figure 5-2. A bounding function is applied repeatedly to an initial circuit behavior interval W^0, producing a sequence of intervals $\{W^0, W^1, W^2, \ldots\}$. As in figure 5-1, each vertical line in the figure represents the set of all circuit behaviors $\mathcal{W}$. In the situation pictured in figure 5-2, if $W^m \subseteq W^p$ for some $m>p$, then it follows from the definition of a bounding function that W^p is an introverted interval. In the figure W^0 is introverted because $W^3 \subseteq W^0$.

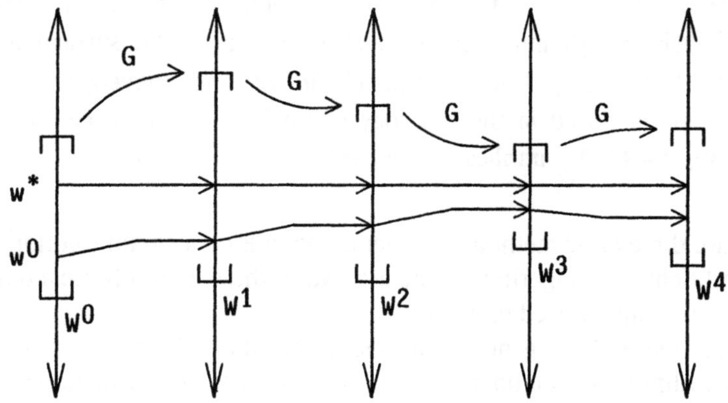

Figure 5-2: The bounding function G is applied repeatedly in theorem 5-2.

Consider any sequence of circuit behaviors generated by the iteration function, starting from a circuit behavior w^p in the introverted interval W^p. Any such sequence must return to the interval W^p at least once every m elements. Therefore, it contains a subsequence that lies completely inside W^p. Since such a circuit behavior sequence converges to w^*, any of its subsequences must as well. Therefore, w^* must lie in the closed interval W^p. In fact, the solution w^* must then lie in all intervals W^i for i>p based on the definition of a bounding function. This reasoning is formalized in the following theorem. Roughly speaking, the solution can be bounded by finding a region of the solution set that "traps" a convergent sequence. Any introverted interval is a guaranteed bound on (i.e., is guaranteed to contain) the circuit solution.

Theorem 5-2: Let $W^0 = [L^0, U^0]$ for any L^0, U^0 such that $L^0 \in \mathcal{W}$, $U^0 \in \mathcal{W}$, and $L^0 \leq U^0$. Let $W^{i+1} = G(W^i)$ $\forall$ integers i≥0. If[62] $W^m \subseteq W^0$ for some $m>0$, then $w^* \in W^i$ for all i≥0.

Proof: Pick any $w^0 \in W^0$ (L^0 for example). Let $w^{i+1} = F(w^i)$ $\forall$ i≥0. Then $w^i \in W^i$ $\forall$ i>0 by the definition of the bounding function G with respect to the iteration function F. Therefore, $w^m \in W^m \subseteq W^0$. But since $w^m \in W^0$, $w^{m+i} \in W^i$ $\forall$ i≥0 by the same reasoning. Similarly, $w^{km+i} \in W^i$ $\forall$ i≥0 and all integers k≥0. For each i, the subsequence $\{w^{km+i}, k=1,2,3,...\}$ lies completely in W^i and converges to w^*. Since W^i is closed in the topology of uniform convergence, i.e., in $L^\infty([0,T]\rightarrow \mathbb{R}^n)$, this implies that $w^* \in W^i$. ∎

There are three basic methods to generate bounds on a circuit solution for use in a bound tightening algorithm. The last two of these methods use a bounding function to find introverted intervals.

The first method to generate a bound on the solution involves direct bounding techniques. If a bound can be very loose, direct calculation can often be applied to even complex circuits. For example, many circuit models are guaranteed to produce voltages that lie within the range of the power supply voltages. The result is a loose bound for the voltage variables in such a circuit. While a bound generated by this method might be very poor, it can serve as an initial bound in a bound tightening algorithm that must start with a rigorous bound. Even if a tightening algorithm needs a rigorous bound to produce guaranteed final bounds, in practice this constraint can be relaxed to produce "bounds" that are very likely to be valid. It is even conceivable that a very conservative timing simulator could generate a good initial bound with a sufficient confidence level.

The second approach involves deriving a circuit behavior interval, based on the circuit being considered, that is guaranteed to be introverted, i.e., contain its image based on some, possibly very simplified, bounding function. Again the

---

[62]No generality is lost by assigning the index of zero to the introverted interval.

bound can be very loose. This method includes either a closed-form solution, or a series of tests on different candidate behavior intervals. Often very conservative estimates for the solution can generate good candidates for introverted intervals. If an interval is found that is introverted when restricted to a portion of the time window [0,T], say [0,T/2], the time step can be partitioned for separate analysis.

The third approach involves the use of a bounding function G that has a convergence property similar to that of the iteration function. In other words, the interval sequence $\{W^0, G(W^0), G(G(W^0)), \ldots\}$ will always converge uniformly to an interval W^* such that $W^* = G(W^*)$, even if the sequence does not telescope inwards. A "fixed point" of a bounding function must be an introverted interval, so carrying such a sequence to convergence will generate a bound on the circuit solution. Recall that calculating a bounding function can be much faster than calculating its corresponding iteration function.

5.2 D.C. Analysis

The first application of bound relaxation considered is d.c. analysis. While it is not the most important application, it provides simple examples of the basic technique. D.C. analysis does not involve variations in time, and thus the complexity of time intervals need not be considered.

5.2.1 Basic Strategy

Even in d.c. analysis, complex circuit models become difficult to bound directly. A convenient way to attack the problem is to use a strategy that is similar to that presented for transient analysis in chapter three. An algorithm can partition the circuit into small blocks and use relaxation methods to combine the results.

Unfortunately, an exact relaxation algorithm for d.c. analysis that is analogous to the transient analysis algorithm does not completely satisfy the requirements for an iteration function given in the last section. A d.c. solution must be a fixed point of a d.c. relaxation function, but there is no guarantee of the existence of a unique solution, let alone a guaranteed convergence property.

A relaxation approach can still be used, though, because only the tightening of bounds is required. The interval [0,VDD] is a guaranteed bound on all the node voltages in any d.c. circuit solution, as long as all the inputs fall within this range. Even if there is not a unique solution, the first property of a bounding function is still valid. Since each solution is a fixed point of the iteration function, relaxing a bound containing all the solutions produces a tighter bound that still contains all the solutions. By relaxing the initial bound [0,VDD], a tighter bound is produced that contains all possible d.c. solutions.

Except in the presence of feedback circuits, d.c. relaxation generally does converge. By analogy with the relaxation discussion in the last chapter, d.c. relaxation converges in each cluster circuit if the cluster contains a resistive path to a fixed voltage source and all transistor gate terminals are fixed inputs. For a combinational logic circuit with no feedback, convergence is propagated from one cluster to the next. Bound relaxation can therefore produce tight bounds in the portion of a circuit that has a unique solution.

In subcircuits without unique d.c. solutions, e.g., latches, the initial bound [0,VDD] can only relax to bounds that include all solutions. Since these solutions generally contain both logic extremes, the bounds remain very loose. In this manner, bound relaxation can help discover portions of a circuit that require more information to produce a unique solution. By shrinking the bound to exclude all but one of the solutions, e.g., forcing some nodes to particular logic values, the relaxation process can then continue towards a tight bound on the desired solution.

5.2.2 Resistor Network Example

The first d.c. example is chosen to illustrate the general procedure of relaxing bounds. A simple resistor circuit is chosen for which, if the resistors are linear, a d.c. relaxation algorithm is guaranteed to converge and satisfy all the requirements of a general iteration function. It demonstrates that the iteration function need not be known precisely to generate a bounding function that may be used to produce a tight bound on the solution. Rigorous bounds can therefore incorporate element uncertainty.

The goal of the first example is to solve for the d.c. voltages in the bridge circuit pictured in figure 5-3. All five resistors are linear[63], with unknown values that lie in the interval between 1Ω and 2Ω. Such a circuit can be solved by an iterative relaxation procedure analogous to Waveform Relaxation. A circuit behavior in this case consists simply of two branch voltages, v_1 and v_2. A circuit behavior here is a vector-valued function with two dimensions defined

---

[63]The resistors are linear here only for simplicity.

on a time interval consisting of a single point. The iteration function F calculates a new value for each branch voltage under the assumption that the other is fixed at its previous guess by an independent voltage source. The resulting closed-form iteration function is

$$\{\hat{v}_1, \hat{v}_2\} = F(\{v_1, v_2\}) =$$
$$\{\frac{R_2R_3(1) + R_1R_2v_2}{R_1R_2 + R_2R_3 + R_1R_3}, \frac{R_5R_3(1) + R_4R_5v_1}{R_4R_5 + R_5R_3 + R_4R_3}\}. \tag{5.1}$$

While such a calculation is normally simple for a circuit containing linear resistors, it is impossible in this example because the exact values of the resistors are unknown.

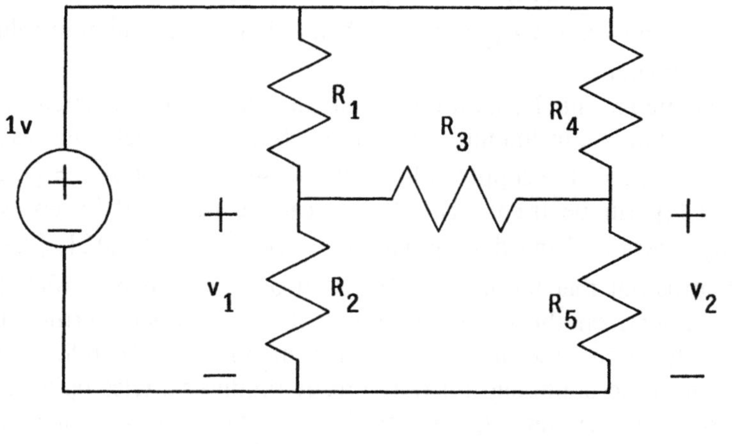

Figure 5-3: Bound relaxation can incorporate element uncertainty.

Even though the iteration function F for the circuit in figure 5-3 cannot be evaluated exactly, it is possible to derive a bounding function G that bounds it. Given bounds on the voltages v_1 and v_2, bounds can be calculated for the voltages $\hat{v}_1$ and $\hat{v}_2$ based on the bounds for the resistor values. Assume that a circuit behavior interval of the form $W = [L, U] = [\{v_{1L}, v_{2L}\}, \{v_{1H}, v_{2H}\}]$ is the input to the bounding function G. Consider only the generation of bounds on the voltage $\hat{v}_1$ since the bounds for $\hat{v}_2$ are completely symmetrical. The expression for $\hat{v}_1$ in terms of R_1, R_2, R_3, and v_2 listed in equation (5.1) is a monotonically increasing function of v_2 and R_2 and a monotonically decreasing function of R_1. It is not monotonic in R_3 so the consideration of R_3 is deferred for now. Therefore, G has the form

$$G([\{v_{1L},v_{2L}\},\{v_{1H},v_{2H}\}]) = [\{\frac{R_3+2v_{2L}}{2+3R_3}, \frac{R_3+2v_{1L}}{2+3R_3}\}, \{\frac{2R_3+2v_{2H}}{2+3R_3}, \frac{2R_3+2v_{1H}}{2+3R_3}\}]. \quad (5.2)$$

The lower bound on $\hat{v}_1$ is a decreasing function of R_3 if $v_{2L} > 1/3$ and an increasing function otherwise. Therefore, the maximum value of R_3 is used only if the input voltage v_{2L} is larger than 1/3. Otherwise, the minimum value $R_3 = 1\Omega$ is used. Similarly for the upper bound, the maximum value of R_3 is used if the input voltage v_{2H} is smaller than 2/3, and the minimum value of R_3 is used if it is larger.

To generate an initial bound on the circuit solution for the circuit in figure 5-3, direct bounding techniques can be used. Since all branch voltages must be within the range of the supply voltage, the loose bound of $W^0 = [L^0, U^0] = [\{0,0\}, \{1,1\}]$ can be used. This bound can be tightened by applying the bounding function G and finding that $W^1 = G(W^0) = [\{0.2,0.2\}, \{0.8,0.8\}] \subseteq W^0$. It turns out that the interval W^0 is introverted, which verifies that it is indeed a bound on the circuit solution. It can be easily verified that the bounding function used here is inclusion monotone. Therefore, repeated application of the bounding function directly produces a telescoping sequence of bounds on the circuit solution. In this case the sequence converges to an interval $W^* = [\{1/3, 1/3\}, \{2/3, 2/3\}] = G(W^*)$. This interval is the tightest possible bound on the circuit solution given the uncertainty in the resistor values, because each endpoint can be obtained by a proper choice of resistor values between 1Ω and 2Ω. Feedback between the two nodes is positive so simplification uncertainty is avoided.

5.2.3 MOS Feedback Circuit Example

A more realistic example of a d.c. MOS circuit model is presented in this subsection. To illustrate a situation with multiple solutions, a latch consisting of two inverters, pictured in figure 5-4, is considered. It is assumed, for the purposes of the example, that the d.c. transfer curves of the two inverters are bounded as pictured in the figure.

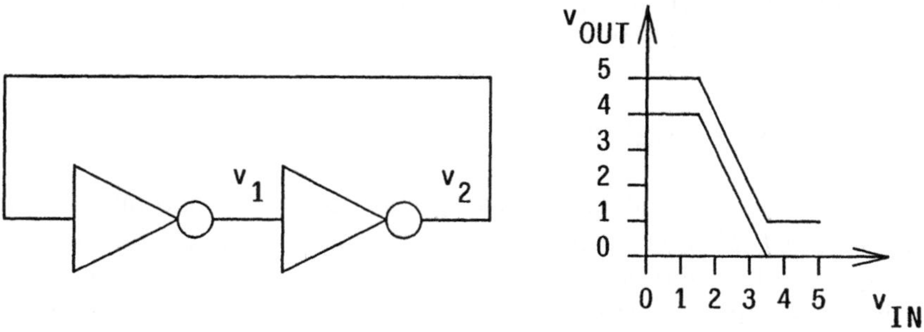

Figure 5-4: Bounds on the d.c. solution of a latch are desired.

As with the last example, a circuit behavior consists of two voltages v_1 and v_2. An initial bound on the solution is $W^0=[L^0,U^0]=[\{0,0\},\{5,5\}]$. This time a Gauss-Seidel form of relaxation is used. If the transfer function of an inverter is denoted $v_{out}=f(v_{in})$, then $\{\hat{v}_1,\hat{v}_2\}=F(\{v_1,v_2\})=\{f(v_2),f(\hat{v}_1)\}$. The bounding function G is defined by the bounds on the transfer function given in the figure.

Evaluation of the bounding function with the initial bound produces $G(W^0)=W^0$. The loose initial bound does not shrink because the relaxed bound must still contain both possible solutions. If a latch has inputs that are tightly bounded, poor bounds inside can pinpoint a problem with multiple solutions specifically to the latch. If the solution is desired in which node v_2 is low, a new initial bound must be chosen in which v_2 is forced low, e.g., $W^0=[\{0,0\},\{5,1.5\}]$.

Evaluation of the bounding function with the new initial bound produces $G(W^0)=W^1=[\{4,0\},\{5,1\}]$. Further relaxation then produces $G(W^1)=W^1$, so W^1 is the tightest achievable bound on the chosen solution. As with the last example, it is also the tightest possible bound given the uncertainty in the devices due to the positive feedback between the two nodes. Even though the d.c. relaxation algorithm is not globally contractive for the latch, it is locally contractive about either solution. Once a bound gets close to a solution, it can be tightened further by relaxation.

Relaxation functions tend to be locally contractive only about stable d.c. solutions. Consider the case of a ring oscillator with the unique solution of all node voltages being at the inverter trigger voltage. Any interval that tightly bounds the solution will relax towards the loose interval corresponding to the supply voltages. Fortunately, the d.c. solution is usually only desired when it is stable, as in the latch example.

5.3 Transient Analysis

In transient analysis, although relaxation of bounds is more complex from an intuitive point of view, theoretically it is simpler due to the guaranteed convergence of waveform relaxation algorithms. A unique transient solution always exists, and relaxation algorithms satisfy the full definition of an iteration function. The added complexity arises from the use of a more complicated iteration function, and the existence of more subtle sources of simplification uncertainty.

An example is provided in this section to illustrate the use of bound relaxation in transient analysis. The simple inverter chain pictured in figure 5-5 is considered. In this example bounds on simple MOS subcircuits are combined to bound a Waveform Relaxation iteration function and generate bounds on a more complex circuit. In this case, to simplify calculations, the bounding function itself is generated with a guess and confirm strategy.

Each inverter in figure 5-5 is modeled with a dependent current source at its output, controlled by both its input and output voltages as illustrated in figure 5-6. All capacitors are assumed to be linear and node voltages are chosen as state variables. The response of the circuit is governed by the system of three differential equations listed below:

$$\begin{aligned}
(C_1 + C_2 + C_3)\,\dot{v}_1 &= f(v_0, v_1) + C_1\dot{v}_0 + C_3\dot{v}_2 \\
(C_3 + C_4 + C_5)\,\dot{v}_2 &= f(v_1, v_2) + C_3\dot{v}_1 + C_5\dot{v}_3 \\
(C_5 + C_6)\,\dot{v}_3 &= f(v_2, v_3) + C_5\dot{v}_2.
\end{aligned} \tag{5.3}$$

Consider an iteration function F_x for the circuit of figure 5-5 based on a waveform relaxation algorithm. The circuit can be decomposed into three subcircuits as pictured in figure 5-7, characterized by

$$C^T{}_i \dot{v}_i = f(v_{i-1}, v_i) + u_i, \quad i=1,2,3. \tag{5.4}$$

The input v_{i-1} represents the input voltage for an inverter while u_i represents coupling current at its output arising from the Miller capacitors. The iteration function,

$$\{\hat{v}_1(t), \hat{v}_2(t), \hat{v}_3(t), \dot{\hat{v}}_1(t), \dot{\hat{v}}_2(t), \dot{\hat{v}}_3(t)\} =$$
$$F_x(\{v_1(t), v_2(t), v_3(t), \dot{v}_1(t), \dot{v}_2(t), \dot{v}_3(t)\}), \tag{5.5}$$

operates on circuit behaviors that consist of six time-varying components, the three node voltages and their derivatives. The iteration function "relaxes" one circuit behavior to another. The three decoupled differential equations that form the iteration function are:

$$\begin{aligned}(C_1 + C_2 + C_3)\,\dot{\hat{v}}_1 &= f(v_0, \hat{v}_1) + C_1\dot{v}_0 + C_3\dot{v}_2 \\ (C_3 + C_4 + C_5)\,\dot{\hat{v}}_2 &= f(v_1, \hat{v}_2) + C_3\dot{v}_1 + C_5\dot{v}_3 \\ (C_5 + C_6)\,\dot{\hat{v}}_3 &= f(v_2, \hat{v}_3) + C_5\dot{v}_2.\end{aligned} \tag{5.6}$$

Each has the form of equation (5.4) and therefore corresponds to the solution of a subcircuit as pictured in figure 5-7. Recall that all signals from other subcircuits (those without hats) are treated as known inputs in the evaluation of each equation in (5.6).

To derive a bounding function G_x for the iteration function F_x specified in equation (5.6), the response of the circuit pictured in figure 5-7 must be bounded. Given bounds on the initial capacitor voltage and the two inputs from the rest of the circuit, G_x must produce bounds for the capacitor voltage waveform and its derivative waveform. There are a number of possible techniques that can be used to bound the response of the logic gate circuit, but for purposes of example, analysis is broken up into small time steps and a table look-up scheme is used to bound the current at the output of the gate. The powerful results derived in sections 4.5 and 4.6 are not used here. It is assumed that the logic gate model produces an output current that is a monotonically

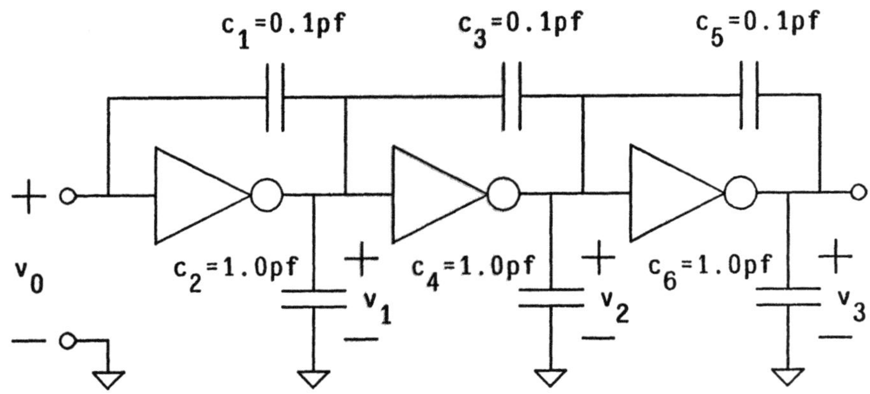

Figure 5-5: The transient analysis of complex VLSI circuits can be partitioned.

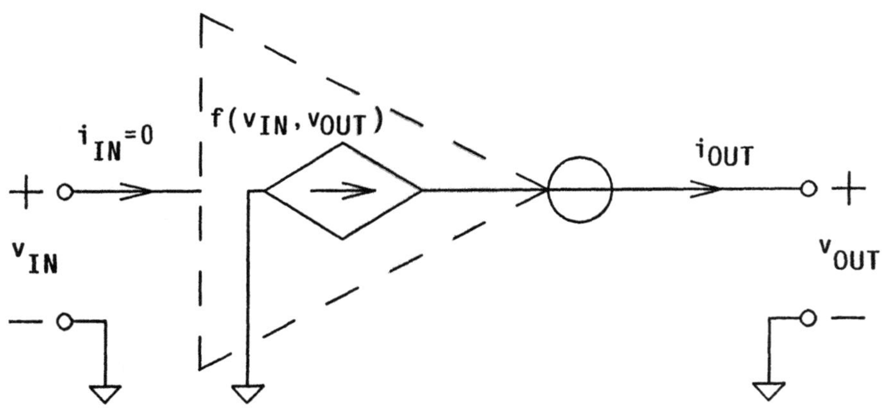

Figure 5-6: The model used for an MOS inverter.

decreasing function of both its input and output voltages[64]. Therefore, with a

[64]For MOS inverter models this is only guaranteed to be true when the voltages fall between GND and VDD. This property is extended here to simplify the example.

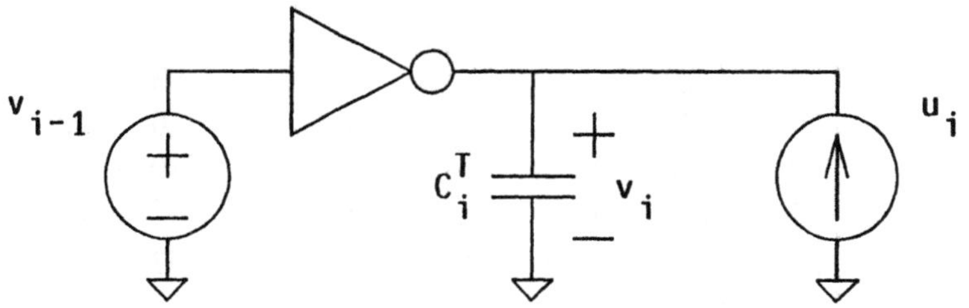

Figure 5-7: G_x produces bounds on the response of each inverter.

table of currents produced for various input and output voltages v_{i-1} and v_i (figure 5-8), bounds on the voltages can be transformed into bounds on the logic gate current function $f(v_{i-1}, v_i)$. Any voltages v_{i-1} and v_i can be rounded up to the next highest table indices to generate a lower bound on $f(v_{i-1}, v_i)$ and rounded down to generate an upper bound. Bounds on the current input u_i are a direct (closed-form) function of bounds on all fixed inputs connected through coupling capacitors. Given a bound on the total current charging the capacitor of figure 5-7, $f(v_{i-1}, v_i) + u_i$, a bound on the derivative of the voltage v_i is produced. Combined with a bound on the initial voltage, this information produces a bound on the voltage over the time interval being analyzed. For example, integrating an upper bound on the derivative waveform starting from an upper bound on the initial voltage produces an upper bound on the voltage waveform.

The main difficulty in bounding the solution of equation (5.4) using the observations outlined above is that bounds on $v_i(t)$ are not generally available initially (except at $t=0$). Bounds on $v_i(t)$ are the desired product, not one of the given inputs. The input interval W can be used to produce an overly conservative bound on v_i for use in calculating $f(v_{i-1}, v_i)$, though, if its use produces an output interval that is a subset of W. Such an output interval is then a valid $G_x(W)$ because the variables cannot escape $G_x(W)$ even when the voltages are assumed to remain inside a larger interval when calculating bounds on their derivatives. This strategy defines what is referred to here as a candidate

$i_{OUT}=f(v_{IN}, v_{OUT})$ (volts vs milliamps)

		v_{OUT}					
		−0.2	0.2	3.0	3.2	3.8	4.0
	0	0.530	0.525	0.448	0.420	0.320	0.280
	3.0	0.080	−0.014	−0.106	−0.107	−0.110	−0.111
v_{IN}	3.2	0.070	−0.030	−0.143	−0.144	−0.147	−0.148
	3.8	0.060	−0.040	−0.253	−0.254	−0.257	−0.258
	4.0	0.058	−0.058	−0.290	−0.291	−0.294	−0.295

Figure 5-8: Partial table of currents for hypothetical CMOS inverter i, i=1,2,3.

bounding function[65].

The analysis is divided into 1 ns time intervals $[t_i,t_{i+1}]$ for i=0,1,..k. A falling step is applied initially to the input of the inverter chain and the particular time interval considered here, $[t_n,t_{n+1}]$, occurs as the output of the second inverter begins to fall. We assume that bounding analysis has already been completed for $t<t_n$. As seen in figure 5-9, at the start of this time step $3.2\leq v_1(t_n)\leq 3.4$, $3.6\leq v_2(t_n)\leq 3.8$, and $-0.1\leq v_3(t_n)\leq 0.1$ (in volts). Based on the behavior in the previous time interval, a conservative guess is made for a bound on the variables in the current time interval. This conservative guess, W^0, is pictured with the outer solid lines in figure 5-9. Bounds on the derivatives are not pictured explicitly in the figure but they are an integral part of the definition of W^0. The interval is based on the guess that $0\leq \dot{v}_1(t)\leq 0.6$, $-0.6\leq \dot{v}_2(t)\leq 0$, and $-0.1\leq \dot{v}_3(t)\leq 0.1$ (in volts/ns). Based on this guess, $3.2\leq v_1(t)\leq 4.0$, $3.0\leq v_2(t)\leq 3.8$, and $-0.2\leq v_3(t)\leq 0.2$ (in volts) are valid bounds on the voltages of W^0 throughout the entire time step $t_n\leq t\leq t_{n+1}$.

To determine if W^0 is an introverted interval and therefore a valid bound on the circuit solution, $G_x(W^0)$ must be calculated. If $G_x(W^0)\subseteq W^0$, then W^0 is introverted. The candidate bounding function just presented can be used

[65] One way to show this result rigorously is to view the candidate bounding function as a bounding function for a Picard iteration function.

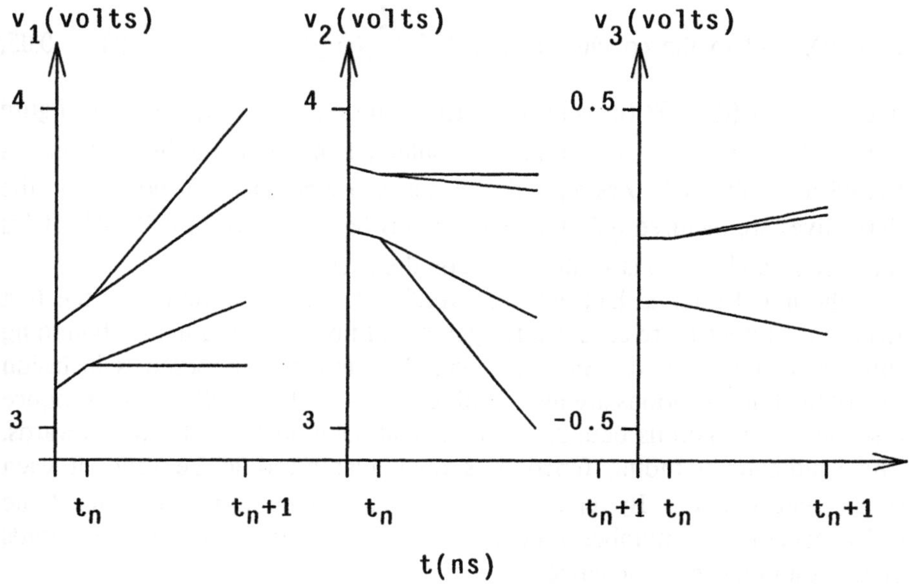

Figure 5-9: Behavior intervals W^0 and $G_x(W^0)$ for the circuit of figure 5-5.

conveniently here because, even though it is not known until later whether it is a valid bounding function, the same condition that guarantees the candidate bounding function is valid is the desired result. The conservative guess W^0 is chosen with the hope that this condition holds.

From the table of values for $f(v_{i-1},v_i)$ in figure 5-8 and the given input interval W^0, it follows that $0.28 \leq f(v_0,\hat{v}_1) \leq 0.42$, $-0.30 \leq f(v_1,\hat{v}_2) \leq -0.14$, and $-0.05 \leq f(v_2,\hat{v}_3) \leq 0.08$ (in mamps). Recall that this calculation of bounds on the inverter output currents uses the guess that the voltages in the solution, $\hat{v}_1$, $\hat{v}_2$, and $\hat{v}_3$, lie within the corresponding voltage intervals of W^0. Bounds on the coupling currents u_i are functions only of bounds on the (known) variables in the initial interval W^0. By straightforward calculation, since addition is a monotonic function, the interval W^0 directly implies that $-0.06 \leq C_3\dot{v}_2 \leq 0$,

$-0.01 \leq C_3\dot{v}_1 + C_5\dot{v}_3 \leq 0.07$, and $-0.06 \leq C_5\dot{v}_2 \leq 0$ (in mamps). These bounds on the currents charging $C^T{}_i$ (both the gate output current and the coupling current), lead to the conclusion that $0.18 \leq \dot{\hat{v}}_1(t) \leq 0.35$, $-0.26 \leq \dot{\hat{v}}_2(t) \leq -0.05$, and $-0.1 \leq \dot{\hat{v}}_3(t) \leq 0.08$ (in volts/ns). The resulting interval (pictured in figure 5-9 as the tighter set of bounds) is contained in the input interval and is therefore a valid $G(W^0)$ as argued previously. Remember that bounds on the derivatives are pictured only implicitly in this figure. Since $G(W^0) \subseteq W^0$, W^0 is introverted and $G(W^0)$ is a valid bound on the circuit solution.

If the initial interval had not been verified to be introverted after the first iteration, another more conservative guess could be tried or a tighter bounding function could be used. Since the bounding function proposed is inclusion monotone (all operations are monotonic), its repeated use will directly produce a sequence of rigorous bounds on the circuit solution that telescope inwards. The candidate bounding function is then guaranteed to be valid at each subsequent iteration. By replacing the bounding function with a tighter[66] one or by increasing the number of iterations used, the accuracy of the final bounds on the solution can be increased.

In the example just presented, the bounding function ignored correlations between a node voltage and itself. A lower bound was calculated assuming that the voltage could be near the upper bound. Because of this simplification, the upper and lower bounds must diverge at each time step. Obviously, this is a fatal simplification in general since it eliminates exploitation of the restoring nature of logic that makes bounding techniques attractive in digital circuits. Therefore, the bounding function just presented is not adequate for efficient bounding, even though it is useful for generating initial bounds. More sophisticated algorithms are presented in the next chapter.

---

[66] Simple examples of methods to tighten G_x include the use of a larger table for the inverter current function, the use of voltage bounds that are nonlinear in time (have a time-varying derivative), or partitioning of the time step.

5.4 Resistive Mesh Circuits

The use of relaxing bounds on a simple special case of the MOS circuit model, linear RC mesh circuits, is considered in this section[67]. There are three main reasons to consider this special case of the general MOS model separately. First, these circuits comprise the standard simple MOS circuit model used in timing analysis programs [58]. They are very important for rough timing analysis whether the RC mesh circuits represent the original model or a circuit whose response bounds that of a more sophisticated model. Second, linear RC mesh circuits have unique properties not possessed by the general models. These properties can be exploited to produce more powerful results with respect to relaxing bounds. Finally, RC mesh circuits provide a simple framework in which to investigate the rough convergence properties of bound relaxation in the general MOS model when partitioning reaches the level of individual node equations.

The RC mesh circuits considered contain only linear capacitors, linear resistors, and constant voltage sources. All of the capacitors and voltage sources must be connected to a common ground terminal, thereby precluding capacitive coupling between two nodes if one is not ground. The voltage sources may have a value of zero, but the capacitors may not. The non-ground terminals of both the capacitors and voltage sources are connected to an arbitrary mesh of resistors. The resistors may not have zero resistance but they may be infinite (open circuits). Any initial condition for the capacitor voltages is permissible.

The RC mesh circuits just described can be used to roughly model a wide range of integrated MOS circuits. Transistors are modeled with a resistor between their drain and source terminals, and with a grounded capacitor at their gate terminal. Integrated wires are distributed RC lines that can be modeled with lumped elements. The operation of a transistor causes its resistance to change from one time interval to another. Initial conditions in the mesh circuits are used to piece the different responses from each time interval together. Figure 5-10 contains an example of an RC mesh circuit along with the MOS circuit it models. In this circuit, the inverter input rises at $t=0$ and the pass transistor turns on at $t=3$ ns. In the time interval from 0 to 3, the resistance of the pass transistor is infinite and the inverter discharges only the voltage v_1. At $t=3$, v_1 has fallen to 2v and the resistance of the pass transistor falls to some finite value. The behavior is then calculated for $t \geq 3$ by solving the RC mesh circuit with the pictured initial conditions at $t=3$.

---

[67] A more complete investigation can be found in [57].

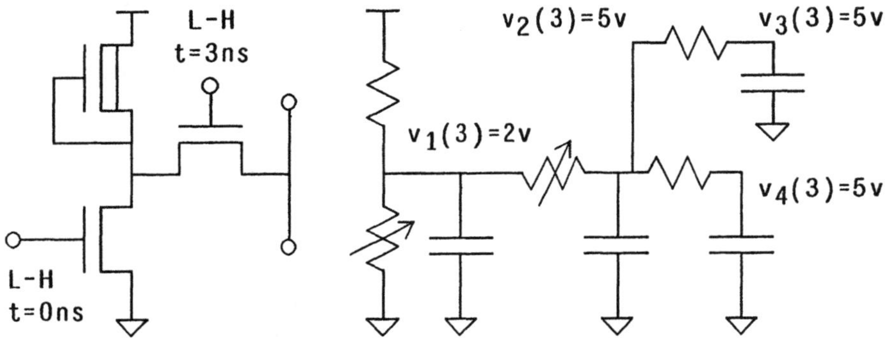

Figure 5-10: Linear RC mesh circuits are often used to model MOS circuits.

Since speed is critical in timing analysis programs, exact solution of the behavior of linear RC meshes is avoided. Simple approximations or bounds are often used. For example, simple bounds have been used for the zero state step response of RC tree circuits [30]. By applying a bound relaxation algorithm to this class of circuits, it is shown in this section that a sequence of closed-form bounds exists that eventually becomes arbitrarily tight. The convergence behavior of this sequence provides a rough estimate of the behavior of relaxing bounds in more general cluster circuits. Although closed-form bounds exist and are currently used for many special RC mesh circuits found in practice, these new bounds based on relaxation are also useful in their own right. They produce a mechanism for tightening bounds to any level of accuracy, and all effects such as charge sharing are accounted for, i.e., they do not place constraints on the initial state of the circuit.

There are several properties unique to RC mesh circuits that can be exploited when relaxing bounds on their solutions. First, the linear nature of the circuits allows closed-form evaluation of the relaxation function for a wide class of initial waveforms. Second, the relaxation function is monotonic, implying that an "optimal" bounding function can be produced by evaluating the relaxation function at each endpoint of the input interval. Third, an initial rough bound always exists that is guaranteed to relax to a tighter bound. Since the sequence of repeated exact relaxations of this bound converges to the solution, telescoping bounds of arbitrary accuracy can be produced by approaching the solution from both sides.

5.4.1 Notation and Definitions

All voltages in the RC mesh circuits considered are normalized to fall between 0 and 1 (volts). In a typical digital VLSI circuit this corresponds to the range of 0 to VDD. The initial conditions and solutions can then be scaled appropriately. Such a simple scaling of voltages is possible due to the linear nature of the circuits considered.

The network $\mathfrak{B}$ represents a general linear RC mesh circuit. $\mathfrak{B}$ contains N positive, linear, grounded capacitors labeled C_I for $I=0$ to $N-1$. A voltage v_I with initial value $v_I(0) = v_{I,0}$ is associated with each capacitor. The nodes are connected to each other and to constant, grounded voltage sources through positive, linear resistor branches. A resistor connecting node I to node J is denoted $R_{I,J}$ or $R_{J,I}$ for $I \neq J$. A resistor connecting node I to ground is called $R_{I,L}$ while one connecting node I to a node with fixed voltage one (typically representing VDD) is called $R_{I,H}$. The finite conductance of each resistor $R_{I,J}$ is denoted $G_{I,J}$. It is assumed that there is a path through resistor branches with positive total conductance between any two nodes[68]. Other voltage sources beyond ground and VDD are not included in the notation because they do not normally occur in digital MOS circuits, but their incorporation is straightforward. A typical node is pictured in figure 5-11.

By applying KCL at each node in $\mathfrak{B}$, each of the N node equations can be derived as follows:

$$C_I \frac{dv_I}{dt} = G_{I,H}(1-v_I) + G_{I,L}(0-v_I) + \sum_{J=0}^{N-1} G_{I,J}(v_J-v_I) \tag{5.7}$$

or

$$\tau_I \frac{dv_I}{dt} + v_I = \frac{G_{I,H}}{G_{I,P}} + \sum_{J=0}^{N-1} \frac{G_{I,J} v_J}{G_{I,P}} = v_{I,S} \tag{5.8}$$

where

[68]If there were two nodes without such a path, the circuit could be partitioned into smaller circuits that could be analyzed separately.

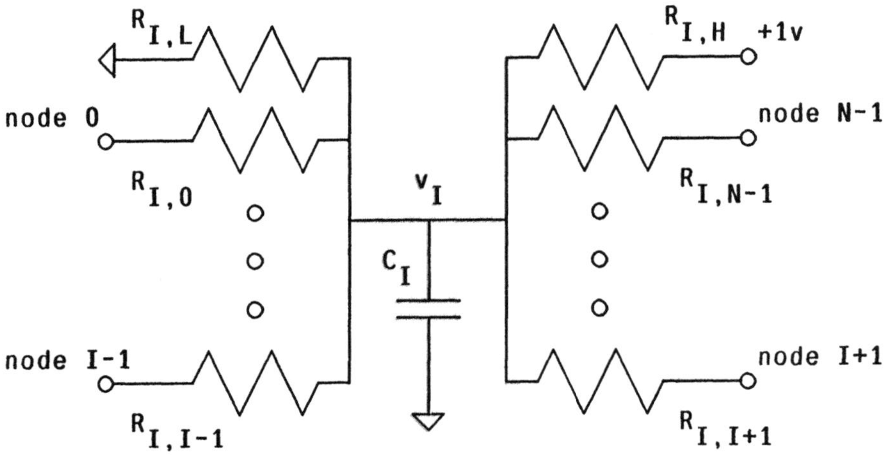

Figure 5-11: A typical node I of the linear RC mesh circuit $\mathfrak{B}$.

$$G_{I,P} = G_{I,L} + G_{I,H} + \sum_{J=0}^{N-1} G_{I,J} \;>\; 0 \text{ and } \tau_I = \frac{C_I}{G_{I,P}} \;. \tag{5.9}$$

The N node equations as expressed in equation (5.8), along with the N initial node voltages, completely determine the behavior of the network for positive time [51]. Each node equation is equivalent to the response of a single pole RC circuit with appropriate input voltage as pictured in figure 5-12. The input voltage, $v_{I,S}(t)$, is a weighted sum of the other node voltages as given in equation (5.8).

As in the more general case, the possible responses for network $\mathfrak{B}$ are called behaviors. An RC mesh behavior consists only of continuous node voltages defined over a finite time interval [0,T]. The solution of the circuit is the unique behavior that satisfies the initial conditions and the node equations (Equation (5.8)).

Definition 5-7: An <u>RC mesh behavior</u> w for network $\mathfrak{B}$ consists of the continuous functions $v_0(t)$, $v_1(t)$, . . . $v_{N-1}(t)$ defined over the time interval $t \in [0,T]$. The <u>RC mesh solution</u> w^* for network $\mathfrak{B}$ is the unique RC mesh behavior such that $v_{I,0} = v_I(0)$ and equation (5.8) is satisfied $\forall I$ such that $0 \leq I \leq N-1$ and $\forall t$ such that $0 \leq t \leq T$.

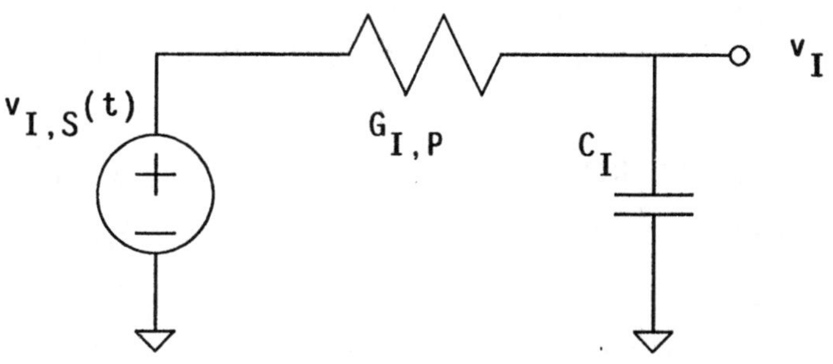

Figure 5-12: Equivalent circuit for node I of network 𝔅.

A bound on the RC mesh solution is defined in the context of the partial ordering of general circuit behaviors defined previously. One RC mesh behavior is said to be greater than or equal to another if its node voltages are never smaller throughout the entire time interval $0 \leq t \leq T$.

Definition 5-8: RC mesh behavior w is said to be greater than or equal to RC mesh behavior $\hat{w}$ if $v_I(t) \geq \hat{v}_I(t)$ $\forall I$ such that $0 \leq I \leq N-1$ and $\forall$ $t \in [0,T]$. Any RC mesh behavior that is less than or equal to the RC mesh solution is said to be a lower bound. Likewise, any RC mesh behavior which is greater than or equal to the RC mesh solution is said to be an upper bound.

The relaxation of an RC mesh behavior produces a new RC mesh behavior by sequentially calculating a new response for each node voltage using equation (5.8), the given initial conditions, and the latest voltage waveform for all other nodes[69]. The relaxation function for network 𝔅 that maps RC mesh behaviors

[69] This represents Gauss-Seidel iteration as opposed to Gauss-Jacobi. Gauss-Seidel is used exclusively in the next chapter for experimental results due to its greater efficiency when implemented on a sequential machine.

to other behaviors is called F. Based on the Waveform Relaxation theorem, as discussed in the last chapter, the sequence $\{w^0, F(w^0), F(F(w^0)), \ldots\}$ converges to the RC mesh solution w^* for any initial RC mesh behavior w^0. The RC mesh solution is also a "fixed point" of the relaxation function, i.e., $w^* = F(w^*)$.

Definition 5-9: RC mesh behavior w is said to relax to $\hat{w}$ ($\hat{w}=F(w)$) if each $v_I(t)$ in $\hat{w}$ solves equation (5.8) with $v_I(0) = v_{I,0}$ and $v_J(t)$ a component of w if J>I and $\hat{w}$ if J<I. The sequence $\{w^0, F(w^0), F(F(w^0)), \ldots\}$ converges uniformly to the RC mesh solution $w^* = F(w^*)$ for any circuit behavior w^0.

5.4.2 Bound Relaxation for RC Mesh Circuits

In this subsection, the three special properties of linear RC mesh circuits mentioned earlier are examined in detail. Their application to bound relaxation is also investigated. It is shown that loose bounds relax (with exact relaxation) eventually to tighter ones. As a result, any bound can be relaxed to one of arbitrary accuracy. Specifically, if a lower bound of zero is used as an initial RC mesh behavior in an exact relaxation procedure, it produces a sequence of lower bounds that converges monotonically to the solution. An identical result holds for upper bounds when using the initial bound of one.

The first special property of linear RC mesh circuits is that the relaxation function can be evaluated in closed form for a wide range of inputs. If an RC mesh behavior has a simple form such as a constant waveform, repeated evaluations of equation (5.8) produce exponentials with polynomial coefficients. The details of such a calculation are presented in the next chapter. Simple equations are derived that calculate new coefficients for each relaxation.

The second special property of an RC mesh circuit is the monotonic nature of each node equation (equation (5.8)). When the voltage waveform $v_I(t)$ is viewed as a function of all the other node voltage waveforms, this function is monotonic in each of the other waveforms. If another node's voltage waveform is replaced by a larger one, the solution for $v_I(t)$ is guaranteed to be replaced by a larger one as well. This property does not depend on the linearity of the circuit elements, only on their individual monotonicity, as demonstrated in chapter four. If another node voltage waveform was replaced by a larger one and $v_I(t)$ was unchanged at a particular time, the current charging capacitor C_I

would be at least as large and as a result, so would the derivative of $v_I(t)$ at that time. The new waveform $v_I(t)$ could not fall below its old value.

In a general circuit the relaxation function is not necessarily monotonic in its input circuit behavior. In RC mesh circuits, though, the relaxation function F consists of a series of solutions for the monotonic node equation (5.8). Monotonicity of the mapping defined by equation (5.8) follows from theorem 4-12 in section 4.5. As a result, the relaxation function F is also a monotonic mapping. This fact is expressed more precisely in the following theorem, which is proven for a more general case in [57].

> **Theorem 5-3:** If w and $\hat{w}$ are any two RC mesh behaviors of network $\mathfrak{B}$ such that $\hat{w} \geq w$, and F is the relaxation function for $\mathfrak{B}$, then $F(\hat{w}) \geq F(w)$.

In terms of the general context discussed in section 5.1, consider the implications of the first two properties of linear RC mesh circuits to the construction of a bounding function. Given an interval $W=[L,U]$, the tightest possible bounding function is given by $G(W)=[F(L),F(U)]$. It follows from theorem 5-3 that this is a valid bounding function and it is the tightest possible because both endpoints of the output interval must be included in the output. Furthermore, because of the first property, this optimal bounding function can be evaluated in closed form.

Because a bounding function exists that is constructed from the exact relaxation function, the results in this section can be derived without appealing to the concept of a bounding function. However, by considering the exact relaxation of a bound in this manner, the relationship between results in the special and general case can be more easily seen. First consider two corollaries of theorem 5-3 with applications to bounding.

The first corollary states that if a bound is relaxed, it is guaranteed to produce another bound. Repeated relaxation of an upper bound will never produce a circuit behavior that falls below the circuit solution w^* at any node and at any time. This is a special case of theorem 5-1 in which the monotonic relaxation function allows upper and lower bounds to be separated. Since exact relaxation of the bound is used, the resulting sequence converges to the circuit solution and repeated relaxation of even a loose bound will eventually produce a bound of arbitrary accuracy.

Corollary 5-4: If w and w^* are two RC mesh behaviors of network $\mathfrak{B}$ such that $w \geq w^*$ and w^* is the RC mesh solution of $\mathfrak{B}$, then $F(w) \geq w^*$. Likewise, if $w \leq w^*$, then $F(w) \leq w^*$.

Proof: Substitute $F(w^*) = w^*$ into the statement of theorem 5-3. ∎

The second corollary is a special case of theorem 5-2, in which upper and lower bounds are separated and the bounding function is guaranteed to be inclusion monotone. It says roughly that if an RC mesh behavior relaxes to one that is larger or smaller, it must be a bound, and its repeated relaxation produces a monotonic sequence of bounds. If an RC mesh behavior is found that is guaranteed to relax in one direction, it will generate a sequence with monotonic convergence.

Corollary 5-5: Let w^0 be an RC mesh behavior with the property that $F(w^0) \geq w^0$. Define w^{i+1} to be $F(w^i)$ for all positive integers i. Then w^0 is a lower bound and $F(w^{i+1}) \geq w^i$ for all positive integers i. Likewise, if $F(w^0) \leq w^0$, w^0 is an upper bound and $F(w^{i+1}) \leq w^i$ for all positive integers i.

Proof: Consider only the first case as the second is completely symmetrical. If $F(w^{i+1}) \geq w^i$ for some i, theorem 5-3 implies that $F(w^{i+2}) \geq w^{i+1}$. Since $F(w^0) \geq w^0$, it then follows by induction that $F(w^{i+1}) \geq w^i$ for all positive integers i. Now assume that w^0 is not a lower bound, so there exists a time t and a component v_I where behavior w^0 is greater than the solution w^* by some amount α. Since the sequence $\{w^0, w^1, \ldots\}$ is monotonic, all elements are greater than the solution by α at time t in component v_I. Since the sequence converges uniformly to w^*, there exists some m such that if i>m, the distance between w^i and w^* for all times and in all components is less than α, leading to a contradiction. Therefore, w^0 is a lower bound. ∎

As a result of the first two properties, all that is necessary to produce a simple telescoping sequence of bounds with arbitrary accuracy is a set of initial bounds with two characteristics. First the bounds must be of a simple form to allow the closed-form solution of the relaxation function. Second, the bounds must relax towards the solution to satisfy the requirements of the second corollary. Fortunately, the third special property of linear RC mesh circuits is that such initial bounds exist, and they are independent of the particular circuit.

The simplest such set of initial bounds is comprised of the constant RC mesh behaviors called L^0 and U^0, with values of 0 and 1 respectively for all nodes and all values of time. Intuitively, since a node will not leave the range of 0 to 1 unless pulled by another that has, no node can escape because none could be first. Therefore, L^0 and U^0 are guaranteed to be valid, though loose, bounds on the circuit solution. In fact, these are the tightest general bounds on an RC mesh circuit because each can be the response of such a circuit. In addition, it is straightforward to show that L^0 and U^0 must relax to tighter bounds [57] and as a result, satisfy the criteria to produce monotonically converging bounds. Other possibilities do exist for initial bounds, such as the closed-form bounds for RC tree circuits mentioned in chapter two. L^0 and U^0, however, are general and simple, and are guaranteed to produce telescoping bounds. They also produce adequate results in many circuits.

Theorem 5-6: Let L^0 be the RC mesh behavior $v_I(t)=0$ $\forall$ $0 \leq I \leq N-1$ and $\forall$ $t \in [0,T]$, and U^0 be the RC mesh behavior $v_I(t)=1$ $\forall$ $0 \leq I \leq N-1$ and $\forall$ $t \in [0,T]$. Then $L^0 \leq w^* \leq U^0$ and the sequences $\{L^0, F(L^0), F(F(L^0)), \ldots\}$ and $\{U^0, F(U^0), F(F(U^0)), \ldots\}$ converge uniformly and monotonically to the RC mesh solution w^*.

Up to this point it has been assumed that the initial conditions of $\mathfrak{B}$ are known exactly. If bounds on the initial conditions are used instead, as when piecing results from different time intervals together, the results in this section can still be used. This follows from the fact that the circuit solution of an RC mesh is a monotonic function of its initial conditions, as discussed for a more general case in section 4.6. If the smaller initial conditions are used with the lower bounds and the larger are used with the upper bounds, the relaxing bounds telescope towards the two circuit solutions associated with the two sets of initial conditions. Since these two solutions are bounds on the actual solution, the relaxing bounds remain valid.

5.4.3 Linear RC Mesh Example

To illustrate the use of bound relaxation on linear RC circuits, the simple circuit pictured in figure 5-13 is analyzed. The circuit has initial conditions of 1 volt at each node. Assume that L^0 is chosen as an initial lower bound over the time interval $0 \leq t \leq T$ for any $T > 0$. Assume all quantities are expressed in units of ns, volts, ma, and KΩ unless otherwise stated.

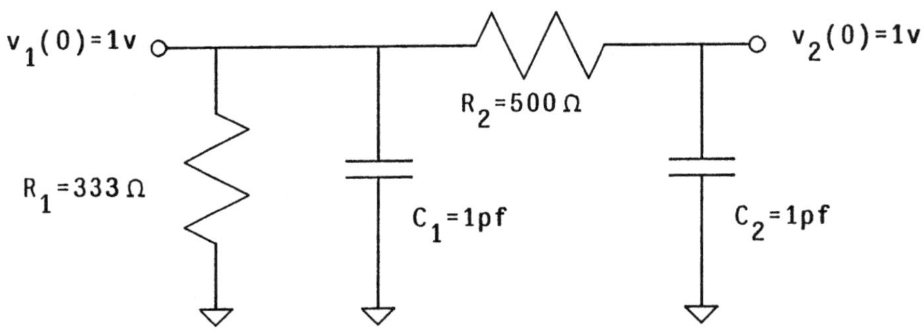

Figure 5-13: Example illustrating bound relaxation for linear RC mesh circuits.

To relax the initial bound, a new voltage waveform must be calculated for each node. The two node equations are $\dot{v}_1 + 5v_1 = 2v_2$ and $\dot{v}_2 + 2v_2 = 2v_1$. The first node is calculated by assuming the second node is fixed at the previous guess, namely zero. The next guess for the first node is then simply e^{-5t}. Note that the second guess satisfies the initial conditions even though the first did not have to. Now the second node response is calculated assuming the first one is fixed at its latest guess, namely e^{-5t}. It can be easily verified that this response is $(5/3)e^{-2t} - (2/3)e^{-5t}$. The relaxed bound consists of the two new voltage waveforms. In the previous notation, $\{e^{-5t}, (5/3)e^{-2t} - (2/3)e^{-5t}\} = F\{0,0\}$.

If the relaxation process is continued by repeatedly relaxing the latest bound, a sequence of lower bounds is generated that approaches arbitrarily close to the solution for all values of time between 0 and T, i.e., for any finite time interval since the choice of T here is arbitrary. For the second relaxation, the lower bound on $v_1(t)$ is the response of node one when driven by the input $(5/3)e^{-2t}-(2/3)e^{-2t}$ at node two. Substitution can verify that this response is $(10/9)e^{-2t}-(1/9)e^{-5t}-(4/3)te^{-5t}$. Continuation of this process produces bounds that approach arbitrarily close to the solution for $v_1(t)$, which for this simple example can be easily calculated to be $(2/5)e^{-6t}+(3/5)e^{-t}$.

5.5 Summary

By extending relaxation algorithms for exact circuit simulation to treat intervals of circuit behaviors, partitioned bounding analysis has been achieved. Examples were provided of how bounding algorithms for small subcircuits can be combined to generate bounding algorithms for large circuits in both d.c. and transient analysis. Relaxation was also applied to linear RC meshes, a special case of the cluster circuit, to generate a sequence of closed-form bounds on their behavior. Experimental results that illustrate the performance of these strategies are presented in the next chapter.

6. ALGORITHMS AND EXPERIMENTAL RESULTS

The high-level bounding strategy presented in the last three chapters gives rise to a wide range of specific bounding algorithms. One goal of this chapter is to illustrate the use of the fairly abstract theoretical results from chapters four and five in constructing a detailed algorithm. The algorithms presented here represent only a small sample of conceivable ones. A second goal of this chapter is to verify earlier claims of efficiency and performance with experimental evidence. Simple bounding algorithms are derived and used to analyze small circuits. The results are used to estimate the potential performance of sophisticated bounding algorithms.

The first section treats only one-way restoring logic circuit models for which specialized bounding techniques are possible. These models can be used for fairly accurate characterization of gate array circuits and are therefore very useful by themselves. Their simplicity also makes them a good starting point for an investigation of more general bounding algorithms.

The second section introduces relaxation algorithms in the simple context of linear RC mesh circuits. These circuits arise in models for interconnect, in very simple approximate circuit models, and in the inner loop of bounding algorithms for more sophisticated models. Relaxation algorithms provide a useful new tool in the analysis of these circuits. In addition, linear RC mesh circuits provide a good introduction for the use of relaxation techniques in more general bounding algorithms.

The third section presents a simple bounding algorithm for the general circuit model presented in chapter four. Experimental circuits are specified with a general list of circuit connections (netlist) similar to that commonly used for SPICE, and partitioning and scheduling are done automatically. Both d.c. and transient analysis are then performed. Uncertainty management, i.e., decisions about accuracy settings and iteration counts for various circuit blocks, is accomplished with user assistance. While the simple bounding algorithm leaves a great deal of room for improvement with many circuit types, it gives an indication of the potential performance of bounding algorithms based on the proposed strategy. In addition, the limitations of the simple algorithm serve to identify important areas for future research.

6.1 One-Way Logic Gate Models

This section deals exclusively with the one-way logic gate model presented in section 4.6.2. This model consists of series-parallel pullup and pulldown transistor groups (restoring logic gates) that are driving RC trees (interconnect). Both Miller capacitance and capacitance inside the transistor groups must be modeled with a single load capacitor.

To further simplify calculations, three additional assumptions are made, the first two of which preclude very accurate modeling of short channel effects. First, the current functions of all transistors within a pullup or pulldown group must be identical except for a multiplicative constant corresponding to their shape factors (width divided by length or "W/L ratio"), i.e., $i_D = S \cdot F(v_G, v_D, v_S, v_B)$ where the function F is identical for all transistors. Second, the current function F must have the property that, when a transistor is split at some point along the channel and analyzed as two separate transistors with identical gate and substrate voltages, the resulting drain current is unaffected. Based on the first two assumptions, any two transistors in series or parallel that have the same gate and substrate voltage are equivalent to a single transistor with an appropriate shape factor. The third assumption restricts all resistors to be linear. This is a good approximation for most interconnect circuits and it allows the direct use of closed-form bounds for linear RC tree circuits.

The derivation of a detailed bounding algorithm for this model is partitioned into steps. Each step represents a transformation to a simpler model with bounding behavior. Only individual logic gates are considered because, due to the one-way nature of the gate models, bounds on the behavior of each gate can be used to propagate bounding information throughout an entire logic circuit. The first subsection discusses transforming arbitrary logic gates into inverters while the next two deal with inverters driving different types of loads. In each case the transformations are very simple to compute, producing bounding algorithms with speed comparable to that of approximating algorithms using simplified models of similar complexity.

6.1.1 General Logic Gates

In terms of the terminology introduced in section 4.6.2, the goal of this subsection is to bound the behavior of an arbitrary *G*-element by that of a simple *G*-element. Instead of using exact models for the pullup and pulldown networks in a simple *L*-element, single transistor models are derived that bound the behavior of the original transistor networks in the context of an *L*-element.

In general, a *G*-element model might contain any number of transistors. From the results derived in chapter four, bounding the output of the *G*-element will lead to a bound on the response of the simple *L*-element in which it is contained. A simple *G*-element with only two transistors always exists, along with a corresponding set of input waveforms, that has an output that bounds the output of the original *G*-element. Roughly speaking, each logic gate can be transformed into a "bounding" inverter.

The approach used to simplify a *G*-element model is to recursively combine the series and parallel transistors in the pullup and pulldown networks into single transistors with bounding properties. The input waveforms on the transistor gates are treated as an integral part of the transistor. At each step, there are two possible simple combination techniques, corresponding to modifying the input waveforms and modifying the transistor characteristics.

First, as shown in section 4.6.2, the output of a *G*-element is a monotonic function of all its input waveforms. As a result, an upper bound on the output of a *G*-element can be obtained by decreasing any or all inputs. Consider any pair of transistors in a pullup or pulldown network that are in series or parallel, with gate voltage waveforms $v_A(t)$ and $v_B(t)$ respectively. If each waveform is replaced by $v_C(t)=\min(v_A(t),v_B(t))$ $\forall t\in[0,T]$, the output of the *G*-element is increased and the two transistors in question have identical voltage waveforms on their gate terminals. The behavior of the pair is then identical to that of a single transistor of shape $S_C=S_A+S_B$ in the parallel case and

$S_C = S_A S_B / (S_A + S_B)$ in the series case[70], with a gate voltage of $v_C(t)$. The use of a maximum voltage will decrease the output of the *G*-element.

Second, also as shown in section 4.6.2, a *G*-element is a monotonic function of all its transistors. It is a monotonically increasing function of all the transistors in the pullup and a monotonically decreasing function of those in the pulldown. A short circuit is an upper bound on a transistor while an open circuit is a lower bound[71], so one transistor may be eliminated in a series or parallel pair. If two transistors are in parallel, either one alone will provide a lower bound on the pair because the second can only increase the current function. Similarly, if two transistors are in series, either one alone will provide an upper bound. If the input waveforms on the two transistors are $v_A(t)$ and $v_B(t)$, either one by itself, applied to its corresponding transistor, provides a valid bound on the transistor pair.

By combining the previous two techniques, reasonable bounds can be obtained for any transistor pair. Each input is restricted to four possible "types" of waveforms: rising, falling, high, and low. Recall that nonmonotonic bounds can be decomposed into rising and falling bounds as mentioned in chapter three. The result is ten separate cases (see figure 6-1 for a list) each for series and parallel combinations. Only the pulldown network is explained in detail as the discussion for CMOS pullup circuits is symmetrical. When discussing the pulldown, recall that an upper bound with respect to the pulldown produces a lower bound with respect to the entire *G*-element. As a result, upper bounds on the gate voltages, while producing a lower bound on the gate output, are said to produce an upper bound on the pulldown.

The first technique is the only option for an upper bound on a parallel combination or a lower bound on a series combination. For n-channel devices the lower bound is produced using the minimum of the two waveforms while the upper bound uses the maximum. In all ten cases of possible input combinations for the upper bound on a parallel combination or the lower bound on a series combination, the first technique provides a reasonable bound,

---

[70] For the parallel case, scale each transistor to unit length and add the widths. For the parallel case, scale each transistor to unit width and add the lengths. The first two assumptions listed at the beginning of this section for the transistor models are needed to justify these transformations.

[71] Strictly speaking, these extremes are not valid transistor models. However, it can easily be shown that a single transistor bounds the pair it is contained in.

i.e., an "on" transistor in series with an "off" transistor is never bounded by an "on" transistor.

v_A	v_B	PARALLEL BOUND	SERIES BOUND
HIGH	HIGH	MINIMUM	MAXIMUM
HIGH	LOW	v_A	v_B
HIGH	RISING	v_A	v_B
HIGH	FALLING	v_A	v_B
LOW	LOW	MINIMUM	MAXIMUM
LOW	RISING	v_B	v_A
LOW	FALLING	v_B	v_A
RISING	RISING	ANY	ANY
RISING	FALLING	v_A AND v_B	v_A AND v_B
FALLING	FALLING	ANY	ANY

Figure 6-1: Lower parallel bounds and upper series bounds (n-channel).

To generate a lower bound on a parallel combination or an upper bound on a series combination, either technique is possible. In each of the ten possible cases pictured in figure 6-1, one of the techniques provides a reasonable bound. In the figure, a single voltage, say v_A, indicates a bound consisting only of the transistor whose gate voltage is v_A. A minimum or maximum of the voltages indicates a bound consisting of a transistor whose shape is derived from those of the original transistors. In many of the ten cases, one of the two techniques clearly provides a more accurate bound. In a few of these cases, the winner is data dependent. In such cases, a choice can be made based on the probability of occurrence, some quick heuristic test can be devised to guess the best choice, or all possibilities can be propagated through the calculation of the gate response. For the case consisting of both a rising and a falling waveform, each waveform must be propagated separately to produce a reasonable bound over the entire simulation time window.

As an example, consider two pulldown transistors of shape 4 that are in parallel as in figure 6-2. First assume that one has a gate voltage that is always low while the other has one that falls low. An upper bound on the current function of the pair can be obtained by taking a maximum of the two upper

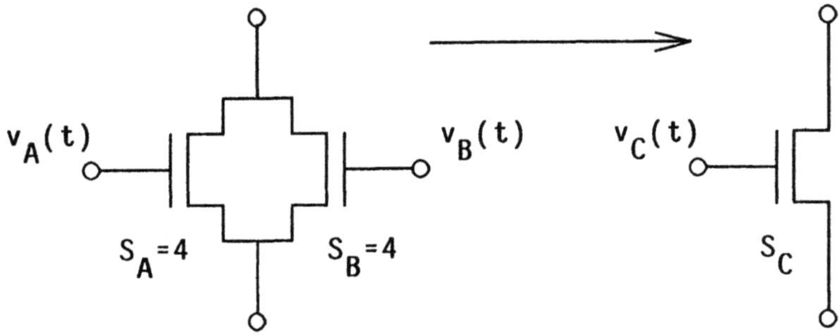

Figure 6-2: Bounding a transistor pair with a single transistor.

bounds on $v_A(t)$ and $v_B(t)$, which in this case is the upper bound on the falling waveform. The shape of the resulting transistor is then 8. A lower bound on the pair is best generated by open circuiting the transistor with the low input. The result is a transistor of shape 4 with the lower bound of the falling waveform as its gate voltage. Next assume that both inputs are rising. A lower bound can be generated by a transistor of shape 8 with a minimum of the two lower bounds for $v_A(t)$ and $v_B(t)$ as its gate voltage or a transistor of shape 4 with either individual lower bound as its gate voltage. If one input arrives well before the other, the first by itself is the best bound. If the two signals arrive at almost exactly the same time, though, the better bound is the transistor of shape 8. Finally assume that one input is rising and the other falling. The best upper bounds are obtained by using either upper bound separately and open circuiting the other transistor. Since each of these bounds is poor in some region (they are monotonic bounds on a nonmonotonic output), they can both be propagated separately. The best lower bound is obtained from the minimum of the two lower bounds. Such a minimum might rise slightly and then fall again, in which case the lower bound of an open circuit for the entire transistor pair might be preferred for reasons of simplicity.

6.1.2 General Inverters

The object of this subsection is to show that the response of simple *G*-elements driving I-elements can be bounded by the zero state response of linear RC trees. Such a response for linear RC trees can then be bounded with closed-form expressions [30]. The three issues that must be addressed are the bounding of the nonlinear time-varying *G*-element output, the bounding of the initial conditions in the I-element, and the bounding of nonlinear capacitors.

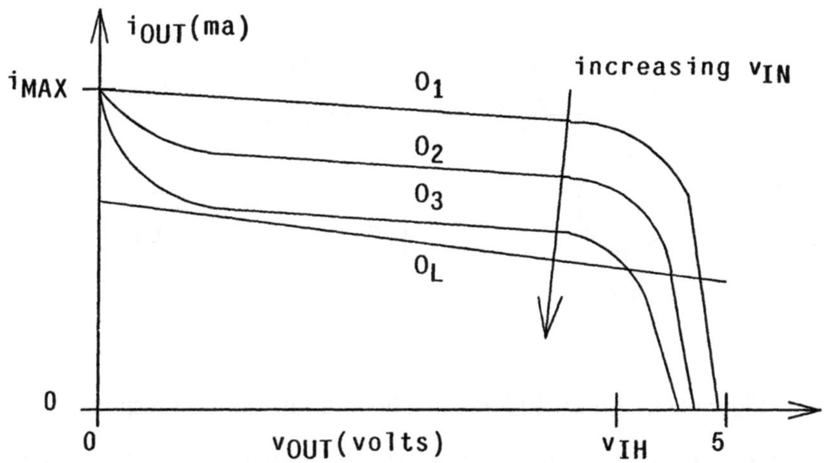

Figure 6-3: Linear lower bounds on a *G*-element output for a staircase input.

The output of a typical simple NMOS inverter is shown in figure 6-3 for the case where staircase[72] input voltage bounds are used. Recall from chapter four that i_{OUT} is the current entering the inverter load while v_{OUT} is the voltage on the node connected to both the load and the driving transistors. Over finite time intervals, the *G*-element output, e.g., O_1, O_2, or O_3, is not time varying. For each such time interval, a linear source can be found that serves as either an

---

[72]Staircase waveforms are not continuous as assumed for all waveforms in chapter four, but any finite number of discontinuities with respect to time can be treated as separate problems over a number of time segments and combined through initial and final states in each segment.

upper or a lower bound on the output. Figure 6-3 pictures a lower bound on output O_3, O_L, that is valid for any value of $v_{OUT} \in [0, V_{IH}]$. As shown in chapter four, the use of such a bound will produce a lower bound on all the outputs of the I-element, as long as the output v_{OUT} stays below v_{IH}. The bound is no longer valid when the output voltage goes above v_{IH}, but, if a restoring logic constraint is used, output voltages are not calculated in this region[73]. Linear bounds can be calculated separately for both the pullup and pulldown network and combined since a *G*-element is monotonic in both networks.

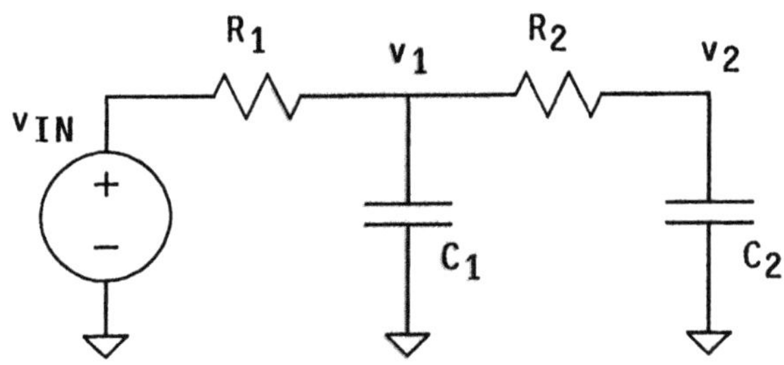

Figure 6-4: Linear RC tree that bounds interconnect response.

There is now a linear source, characterized by a voltage source in series with a linear resistor, driving an RC tree with possibly nonlinear capacitors. Since a zero state response calculation is ultimately desired, uniform initial conditions throughout the interconnect network are needed. To obtain a lower bound on the initial state of all the capacitors in a given time interval, a minimum of all the final voltages in the previous time interval may be used. Likewise, a maximum may be used for an upper bound. The initial state used in the first interval can be the d.c. solution of the circuit with its initial inputs, if the circuit is initially at rest. Otherwise, the initial values must be specified along with the input waveforms. The result of bounding the initial conditions will also bound the response of the *L*-element as derived in chapter four.

The final step in generating a linear RC tree is to replace the capacitors by

---

[73] This is only true for the voltage v_{OUT}. When calculating the bounds for other voltages in an I-element, any voltage is possible in general for v_{OUT}.

linear ones that bound the ones in the basic model. The input to the I-element is no longer time-varying over a given time interval, so it is a monotonic input as required by theorem 4-15. The initial state must be consistent with either a rising or falling input depending on whether the initial node voltages are greater than or less than the source voltage. In chapter four it was shown that larger capacitors will produce an upper bound on the outputs if the voltage source is lower than the initial node voltages and a lower bound if the source voltage is higher. Once the capacitors are linear, superposition can be used to transform the problem to the zero state step response of a linear RC tree as pictured in figure 6-4.

6.1.3 Inverters with Capacitive Loads

This section considers the simple and common case where the I-element loading a logic gate is a single capacitor. The goal is to show that in such a case, one can use exact solutions for various inverters driving capacitors to bound the response of the circuit. These inverter responses are calculated before a simulation is run and can provide very tight bounds on the response of simple *G*-elements.

Before methods for precalculation can be considered, it must be shown that an arbitrary simple *G*-element can be bounded by a "standard" inverter used in the precalculation. If an arbitrary *G*-element is bounded by a *G*-element that contains only two transistors, the transistors will often have a ratio of shapes that is close to some standard for a given process. Logic gates are usually designed so that the strength of their pullup and pulldown networks maintain roughly the same ratio throughout a circuit. Whatever the shapes of the two transistors in the inverter *G*-element are, by adjusting the shape of one of them, a bound can be constructed that has one of a standard set of shape ratios.

As an example consider an NMOS inverter with $S_{pu}=1$ and $S_{pd}=4.5$. Assume that the standard precalculated pullup to pulldown shape ratios are 1:3, 1:4, 1:5, and 1:6. A lower bound on a *G*-element can be produced either by increasing the shape factor of the pulldown or reducing the shape factor of the pullup. Therefore, a lower bound on the example inverter could be one with shapes 1 and 5 or 0.9 and 4.5, both of which have a standard ratio of 1:5. If the output is rising, the best lower bound is the one with shapes 1 and 5, because it has the largest drive, and if the output is falling the other inverter, with shapes 0.9 and 4.5, provides the better bound. In a similar manner, an upper bound on the example inverter is one with shapes 1 and 4 or 1.125 and 4.5, both of which have a standard ratio of 1:4.

Once an inverter of a standard ratio is used, its solution is merely a scaled version of the solution for a standard inverter with the same ratio. The time axis of the solution can be scaled to account for different ratios of transistor shapes to load capacitance. Such a method is compatible with nonlinear capacitors as long as the capacitors, as well as the transistors, have the same form except for a constant throughout the circuit.

The first of two methods for precalculation involves selecting a few "typical" input waveforms. Since standard inverters can be used to bound a *G*-element, the response of these standard inverters driving capacitors can be calculated once for each new process with a set of typical waveforms and the results stored in a table for reference. One of the typical input waveforms can be shifted in time to form a tight bound on the actual input waveform as pictured in figure 6-4. The response to this typical input is then a bound on the original response.

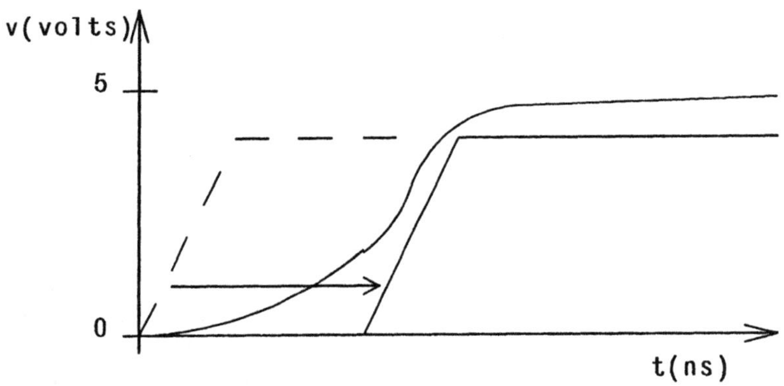

Figure 6-5: Using the response to a "typical" input waveform.

The second precalculation method involves building a bound from a number of precalculated trajectories. When a one-way model for an inverter is driving a capacitor, the equation governing its behavior is

$$i_{OUT}(v_{OUT},t)=C(v_{OUT})\frac{dv_{OUT}}{dt}, \tag{6.1}$$

leading to the solution

$$v_{OUT}(t) - v_{OUT}(t_0) = \int_{t_0}^{t} \frac{i_{OUT}(v_{OUT},\tau)}{C(v_{OUT})}\, d\tau. \tag{6.2}$$

If functions $\hat{i}(v_{OUT})$ and k(t) can be found such that $i(v_{OUT},t) \leq k(t)\hat{i}(v_{OUT})$ or $i(v_{OUT},t) \geq k(t)\hat{i}(v_{OUT})$, a bound on the solution can be found by solving the equation

$$\int_{t_0}^{t} k(\tau)d\tau = \int_{v_0}^{v} \frac{dv_{OUT}}{\hat{i}_{OUT}(v_{OUT})C(v_{OUT})}. \tag{6.3}$$

If a nonlinear load capacitor is used the form of the nonlinearity must always be the same, as only a multiplicative constant can be removed from the precalculation integral to be specied at run time. The bounding of $i_{OUT}(v_{OUT},\tau)$ by a separable function can be quite tight if it is done piecewise, i.e., different functions are used over different time intervals.

When staircase bounds are used for the input voltage waveform, during each time-interval (or step) the input is constant, so the current function is time-invariant (k(t)=1) and this method can be used exactly. The response can be precalculated with each possible step voltage on the input. During each time interval where the input is constant, the precalculated trajectory for that input voltage can be used. Figure 6-6 shows the trajectory of output current and voltage for a rising staircase input with four steps. For each of the load lines corresponding to the four input voltages, a precalculated table can be used to determine the amount the output voltage will fall in a given amount of time, with a given initial voltage.

6.1.4 Experiments

An experimental program has been written that implements some of the transformations discussed in this section. Waveforms are characterized by a list containing a waveform type, e.g., rising or falling, a bound type, e.g., upper or lower, and the times at which it is guaranteed to reach a sequence of threshold voltages. For the NMOS combinational logic circuits tested, two threshold voltages were used for falling waveforms and five were used for rising ones.

For each logic gate, output voltage waveforms are calculated based on the gate model and its input voltage waveforms. Calculated signals are propagated

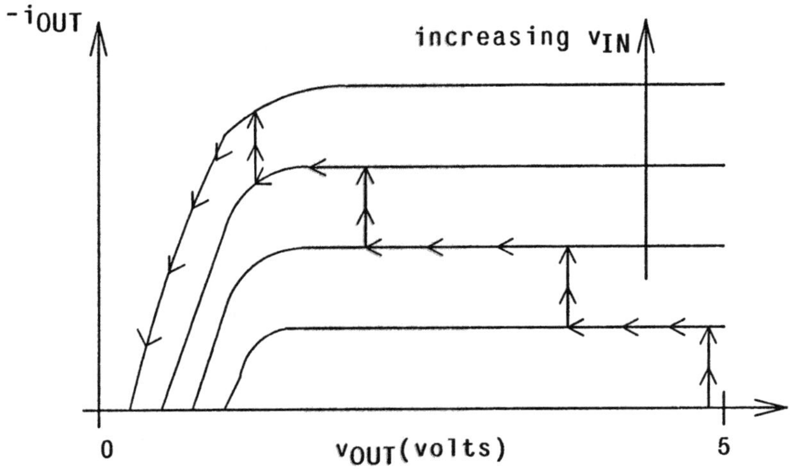

Figure 6-6: Integrating a staircase input.

sequentially[74] through a circuit from inputs to outputs. For a given set of simplifications, the percentage uncertainty in delay remains roughly constant throughout a restoring logic circuit. Assume that the average extra uncertainty is α ns per gate and the average delay of a gate is β ns, resulting in a fractional uncertainty of α/β for each gate. After n stages, the uncertainty is $n\alpha$ ns but the fractional uncertainty is $n\alpha/n\beta = \alpha/\beta$. As a result, only small circuits need be analyzed to get an indication of the fractional uncertainty created by each simplification.

When considering a particular logic gate, the bounding algorithm first transforms the gate into an inverter corresponding to a slow bound. Only the worst case long delay is considered for simplicity, so for nonmonotonic (in time) waveforms, a single set of monotonic bounds is used that is good only for times after the signal settles. The transformation is done recursively using a lookup table containing the information in figure 6-1. Reasonable fixed choices are used for the entries in the table for which more than one choice might be best.

---

[74]Parallel computation is possible for gates whose output waveforms do not effect each other's input waveforms.

The general inverter algorithm is used, but only linear capacitive loads are implemented. For each input voltage, the linear bounds to be used for the pullup and pulldown transistors are stored in a table. After scaling for transistor sizes, a source voltage and resistance are generated for each transistor, (v_1,R_1) for the pullup and (v_2,R_2) for the pulldown. These are then combined into a single source (v_s,R_s) that represents the entire *G*-element through the equations:

$$v_s=\frac{R_2v_1+R_1v_2}{R_1+R_2} \quad , \quad R_s=\frac{R_1R_2}{R_1+R_2} \quad . \tag{6.4}$$

During the time interval over which the input is constant, the output bound is calculated as the response of the linear RC circuit constructed by connecting the load capacitor to the linear source. If the output voltage is v_0 at time t_0, the starting time for the interval, the output voltage v at any later time t is given by:

$$v=v_s+(v_0-v_s)e^{\frac{t_0-t}{R_sC}} \quad . \tag{6.5}$$

The output waveform must be transformed into a staircase bound for use as an input to the next stage, so it is also necessary to calculate the time at which threshold voltages are reached with the equation:

$$t=t_0+R_sC\ln\frac{v_0-v_s}{v-v_s} \quad . \tag{6.6}$$

Figure 6-7 contains an example that illustrates the amount of additional uncertainty introduced by the simplifications used in the implemented bounding algorithm. A chain of four NMOS NAND gates driving linear capacitive loads[75] is considered. A lower bound on the input is given as a rising step function.

The first curve in figure 6-7 represents the exact response of the circuit to the step input as calculated by SPICE. This is the most accurate possible lower bound on the output waveform that can be produced from the given information. The second curve represents the reduction of each NAND gate to a bounding inverter. Approximately 10% uncertainty is introduced in the delay

---

[75] If resistive interconnect models arise, further uncertainty is introduced in bounding their response with the algorithms discussed in subsection 6.1.2 for general inverters.

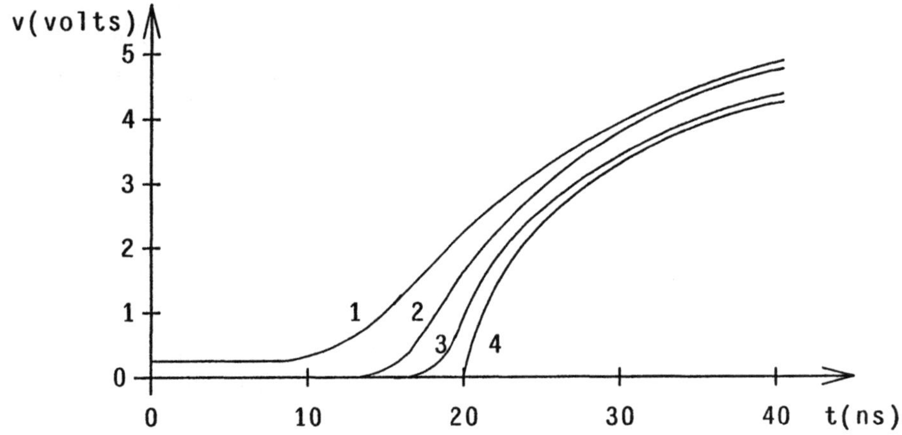

Figure 6-7: Lower bounds on the output of a chain of 4 NAND gates.

of the whole circuit. The third curve represents the reduction of all internal waveforms to staircase bounds with two steps for falling waveforms and five steps for rising ones. This elimination of continuous time variation in the gate drives introduces roughly another 10% delay uncertainty. The final curve represents the use of linear models for the transistor portions of the gates. Only an additional few percent of delay error is introduced by this transfer to piecewise exponential output waveforms. The experiment illustrates that very simple bounding algorithms using linear device models achieve "accuracy" for logic gates that is close to that achieved by similar approximating algorithms [17], without expending a great deal more in computation costs.

6.2 Linear RC Mesh Circuits

This section deals exclusively with the linear RC mesh model presented in section 5.4. This model consists of grounded linear capacitors and grounded constant voltage sources connected through a mesh of linear resistors. Any initial conditions are allowed for the capacitors. A detailed bounding algorithm is derived for this model based on the relaxation strategy developed in section 5.4. An implementation of the algorithm in APL is presented along with some examples of its use. One example illustrates how the algorithm handles charge sharing effects not considered in conventional RC bounding programs.

6.2.1 Bounding Algorithm

In this section an algorithm is presented that calculates relaxations of RC mesh circuit behaviors, including L^0 and U^0, in closed form. The relaxation function consists entirely of solutions to the simple node equation (equation (5.8)). As shown in chapter five, a relaxation function can be used to tighten initial bounds on the circuit solution.

Consider first the form that the circuit behaviors take. Each node equation has associated with it a particular time constant τ_I, called a local time constant. The local time constant of each node is the node capacitance multiplied by a parallel combination of all the resistance terminating on the node (equation (5.9)). The circuit behaviors must contain exponential terms with each of the local time constants because time constants move through the network during successive relaxations. In addition, since these exponentials appear in the function $v_{I,S}(t)$ of their associated node, the coefficients to these exponentials become growing polynomials. Of course, the circuit behaviors must also contain constant terms and exponential terms with any time constants contained in the initial circuit behavior. Any additional time constants cannot arise in the solutions of equation (5.8).

The general form of each node voltage waveform is:

$$
\begin{aligned}
v_I(t) = {} & a_{I00} \\
& + (a_{I10} + a_{I11}t^1 + \ldots + a_{I1M}t^M)\, e^{t/\tau_0} \\
& + (a_{I20} + a_{I21}t^1 + \ldots + a_{I2M}t^M)\, e^{t/\tau_1} \\
& \quad \cdot \\
& \quad \cdot \\
& + (a_{IG0} + a_{IG1}t^1 + \ldots + a_{IGM}t^M)\, e^{t/\tau_G}.
\end{aligned}
$$

The first n time constants are the local time constants determined by the circuit, and the remainder are those supplied by the initial circuit behavior. This can be expressed by a matrix of coefficients A_I for each node I. The components of A_I are denoted $a_{I,K+1,L}$ where K is the time constant and L is the order of the polynomial term. The terms $a_{I,0,L}$ for $L \neq 0$ are zero. The number of columns in the matrix grows as the circuit behavior becomes more accurate through further relaxations. The number of rows remains constant at one more than the number of nodes plus the number of time constants in the initial circuit behavior.

The process of relaxing a circuit behavior involves calculating new solutions to equation (5.8) where the driving voltages are of the form just presented[76]. The solution of equation (5.8) involves two steps: calculating the driving function $v_{I,S}$ and then solving the first order differential equation. The first step is straightforward because it involves taking a weighted sum of the other node voltage waveforms. This translates into a term by term weighting of the matrix coefficients. In matrix notation this becomes:

$$A_{I,S} = \frac{G_{I,H}}{G_{I,P}} \begin{bmatrix} 1 & 0 & . \\ 0 & 0 & . \\ & \cdots & \end{bmatrix} + \sum_{J=1}^{N} \frac{G_{I,J} A_J}{G_{I,P}} \quad . \tag{6.7}$$

The second step, solving the differential equation, can best be accomplished by considering each row of the driving matrix $A_{I,S}$ separately and adding the results. The first row (K=0) is merely a constant $a_{I,0,0}$ that produces an identical constant for its component of the particular solution. All other rows have the form:

$$v_{I,S}(t) = \left(\sum_{L=0}^{M} a_{I,K,L}\, t^L \right) e^{-t/\tau_{K-1}} \quad . \tag{6.8}$$

For this case the corresponding component of the solution has the form:

$$v_I(t) = \left(\sum_{L=0}^{M} \hat{a}_{I,K,L}\, t^L \right) e^{-t/\tau_{K-1}} \quad . \tag{6.9}$$

Substituting these into equation (5.8) implies that

$$(1 - \tau_I/\tau_{K-1})\hat{a}_{I,K,L} = a_{I,K,L} - \tau_I(L+1)\hat{a}_{I,K,L+1}$$
$$\text{where } \hat{a}_{I,K,L+1} = a_{I,K,L} = 0 \text{ if } L>M. \tag{6.10}$$

For a row where $\tau_{K-1} = \tau_I$, this reduces to

---

[76] Laplace transforms can also be used to derive relaxation equations for linear circuits. Such a presentation that uses a Gauss-Jacobi algorithm in a more general context can be found in [59].

$$\hat{a}_{I,K,L} = \frac{a_{I,K,L-1}}{L\tau_I} \quad \text{for } L \geq 1 . \tag{6.11}$$

Recall that there is at least one row which will match each local time constant. Therefore, since $\hat{a}_{I,K,0}$ is unrestricted in this case, it always provides the necessary degree of freedom to satisfy the initial condition. If the local time constants are not unique and two exponential terms exist with the same local time constant, any one can be used to satisfy the initial conditions. The initial voltage $v_{I,0}$ is the sum of all the elements in the first column of A_I.

An APL program that implements the relaxation algorithm for linear RC mesh circuits is listed in figure 6-8. The main program, called SHRINK, takes as arguments both an iteration counter and a three dimensional initial circuit behavior matrix $(A_0, \ldots A_{N-1})$. It returns a list of the relaxed circuit behaviors after each iteration in a four dimensional matrix. Within each iteration loop, a node loop calculates the solution to equation (5.8) for each node. To calculate each new voltage waveform, a row loop steps through the particular solutions to each row with equations (6.10) and (6.11). The program EVAL calculates voltage waveform values at a particular time T. It assumes that the last two dimensions of the input GUESS comprise voltage coefficient matrices, and these dimensions are collapsed into a single voltage value.

6.2.2 Experiments

As a first example consider the RC line pictured in figure 6-9. This could represent a model of an interconnect wire being pulled high by a logic gate. Since the initial conditions of all the nodes are zero, the solution is a zero state step response of an RC tree for which fairly tight closed-form bounds exist [30]. The nodes are numbered from left to right to improve the convergence rate, as this is the direction that the signal is flowing. The sequential nature of the calculation of the relaxation function, i.e., the use of a Gauss-Seidel algorithm, allows signals from nodes with lower indices to more quickly influence those with larger indices. Figure 6-10 shows the results of the first four relaxations of L^0 and U^0 bounds as calculated by the program SHRINK for the rightmost node. Each successive bound obtains good accuracy for larger times.

Since the circuit in figure 6-9 can be bounded in closed form with simple exponentials, these can also be used as initial bounds in a relaxation. As derived in [30], a lower bound on the rightmost node is $1-e^{(13-6t)/66}$. Since voltage is monotonic with respect to position in the step response of an RC line, this is a

```
TAU[I]      = τ_I for 0≤I≤G
INIT[I]     = Initial voltage for node I, 0≤I≤N−1
CONST[I]    = Constant term G_I,H/G_I,P for node I, 0≤I≤N−1
RNET[I;J]   = Ratio G_I,J/G_I,P for 0≤I, J≤N−1, I≠J; 0 for I=J

     ∇V←GS EVAL T;ROWS
[1]  ROWS←+/GS×(ρGS)ρT*ι¯1↑ρGS
[2]  V←+/((ρROWS)ρ(1,*−T÷TAU))×ROWS   ∇

     ∇Z←N SHRINK GS;I;K;L;VS;MARK
[1]  Z←(0,ρGS)ρ0
[2]  ITERLOOP: I←0
[3]  NODELOOP: GS←((ρGS)+0 0 1)↑GS
[4]  VS←((1↓ρGS)↑CONST[I])+ +/[0](⍉(ΦρGS)ρRNET[I;])×GS
[5]  GS[I;K;]←VS[K←0;]
[6]  ROWLOOP: →((1↑ρVS)≤K←K+1)/DONE
[7]  →(TAU[K−1]≠TAU[I])/DIFFERENT
[8]  GS[I;MARK←K;]←0,¯1↓VS[K;]÷TAU[I]×1+ι1↓ρVS
[9]  →ROWLOOP
[10] DIFFERENT: L←¯1+1↓ρVS
[11] RATIO←÷1−TAU[I]÷TAU[K−1]
[12] COLUMNLOOP: →(0>L←L−1)/ROWLOOP
[13] GS[I;K;L]←RATIO×VS[K;L]−GS[I;K;L+1]×TAU[I]×L+1
[14] →COLUMNLOOP
[15] DONE: GS[I;MARK;0]←INIT[I]− +/GS[I;;0]
[16] →(0≠+/GS[I;;¯1+1↓ρVS]≠0)/SKIP
[17] GS←((ρGS)−0 0 1)↑GS
[18] SKIP: →((ρGS)[0]>I←I+1)/NODELOOP
[19] Z←(((¯1↓ρZ),¯1↑ρGS)↑Z),[0]GS
[20] →(0<N←N−1)/ITERLOOP   ∇
```

Figure 6-8: APL global variables and programs used in experiments.

bound on the other node voltages as well. Likewise, an upper bound on the leftmost node, $1-(3/11)e^{-t/3}$, is an upper bound on all the node voltages. These bounds, called $\hat{L}$ and $\hat{U}$ respectively, and their first three relaxations are pictured in figure 6-11. Even though these bounds are not guaranteed to produce monotonic convergence, they nevertheless constantly improve in this

```
TAU    = 0.75   1.2    2
INIT   =    0     0    0
CONST  = 0.75     0    0

RNET   = ⎡0.00 0.25 0.00⎤
         ⎢0.40 0.00 0.60⎥
         ⎣0.00 1.00 0.00⎦
```

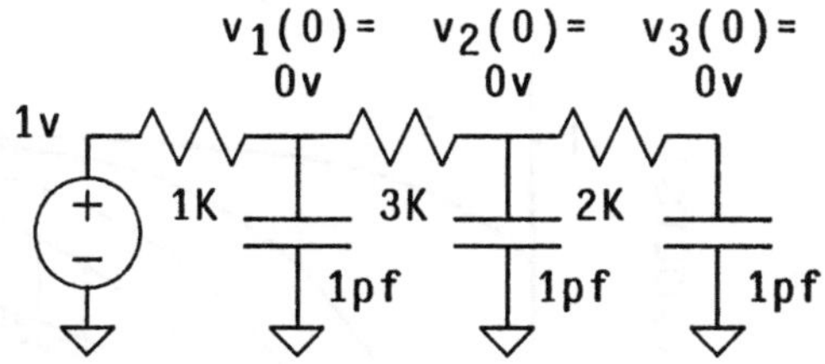

Figure 6-9: The RC line considered as the first example.

example. Since the upper bound $\hat{U}$ is tighter than U^0, the monotonicity of the relaxation function does guarantee that each relaxation produces a tighter upper bound than the corresponding one in figure 6-10. While these initial bounds do not produce faster convergence rates, they provide better bounds for large values of time throughout.

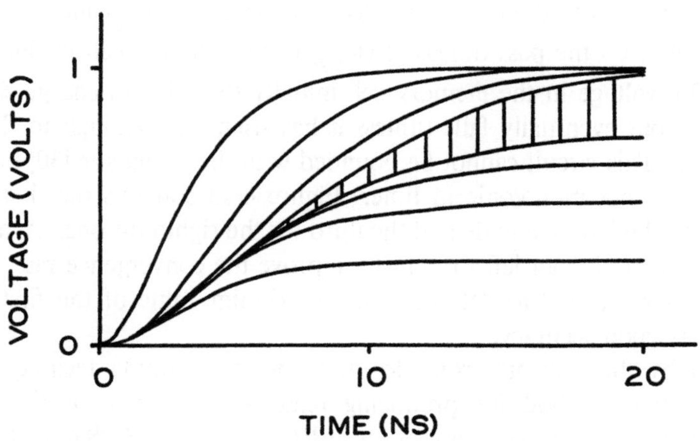

Figure 6-10: Three relaxations of L^0 and U^0.

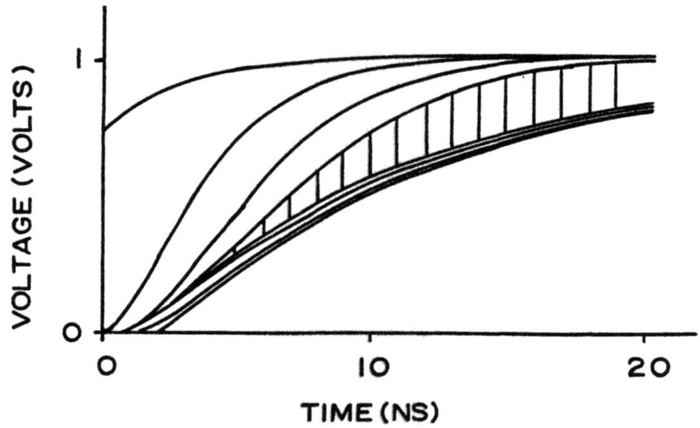

Figure 6-11: Three relaxations of $\hat{L}$ and $\hat{U}$.

As a second example consider the circuit pictured in figure 6-12. This could represent a memory cell discharging a large precharged bus (C_2) and sense-amp input (C_3). At t=0 the pass transistor ($R_{1,2}$) connecting the cell to the bus is enabled. The voltage in the memory cell initially rises due to charge sharing from the bus but eventually falls (unless it has risen high enough to flip the memory cell). This circuit cannot be bounded with the bounds in [30] because the voltages are not monotonic in time. Figure 6-13 contains the first three relaxations for the leftmost node and the third for the rightmost one. Again the nodes are numbered from left to right to improve the convergence rate. Only one relaxation is needed to determine the maximum value of the first node voltage to very good accuracy.

As shown by the previous examples, the bound relaxation technique can provide a simple method for producing reasonable bounds, or tightening unsatisfactory ones, on the behavior of linear RC models for MOS circuits. The polynomial coefficients of the local time constants provide improving

TAU =	0.5	5	1
INIT =	0	1	1
CONST =	0	0	0

$$\text{RNET} = \begin{bmatrix} 0.00 & 0.50 & 0.00 \\ 0.50 & 0.00 & 0.50 \\ 0.00 & 1.00 & 0.00 \end{bmatrix}$$

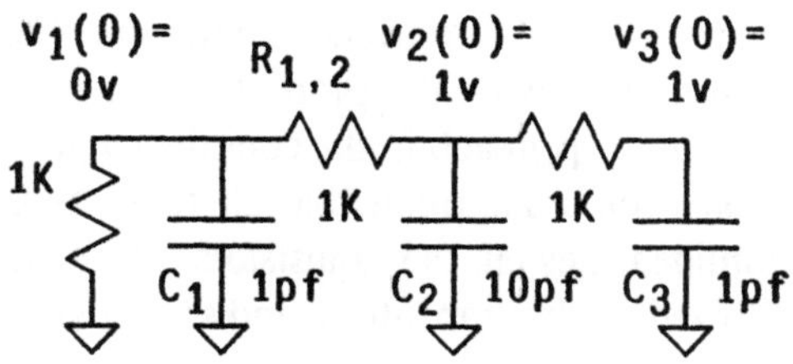

Figure 6-12: The bus driver considered as the second example.

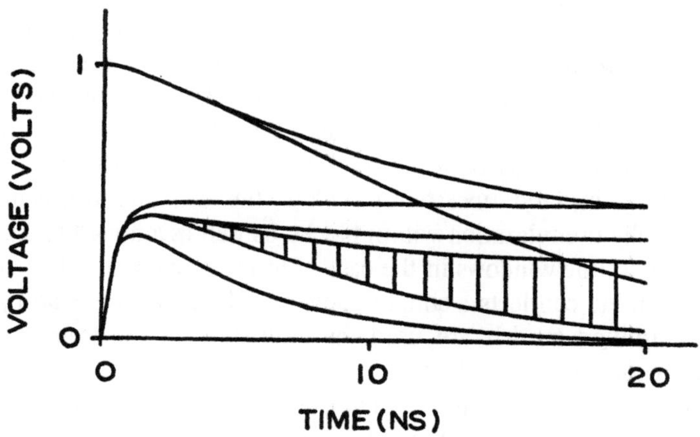

Figure 6-13: Relaxations of L^0 and U^0 for nodes 1 and 3.

correction[77] for the differences between the initial time constants and the eigenvalues of the system. The experiments show that even within cluster circuits, relaxation can exhibit reasonable performance.

---

[77] If a high degree of accuracy is needed, the calculations for each correction grow with order N^2T^2 where N is the number of nodes and T is the number of relaxations.

6.3 General Circuit Models

This section considers the general MOS integrated circuit model presented in chapter four. Restrictions are made to simplify the bounding algorithms without removing the properties essential for demonstrating the feasibility of a bounding approach in the completely general case. The implemented algorithm makes the assumption that all resistors and capacitors are linear, but the key nonlinearities in the transistors are maintained. Device uncertainty is only allowed in the transistors, and is modeled with a setting for extra conservatism in device simplifications. Arbitrary topologies are allowed, as long as each internal node of the circuit has a nonzero grounded capacitance. As a result, simplifications that are possible when some nodes can be internalized into resistive "one-port" groups are not implemented.

The first four subsections describe different portions of the bounding algorithm. In each case very simple detailed strategies are used. The partitioning of the circuit into blocks for relaxation is completely static, i.e., fixed throughout the simulation of a particular circuit, and only simple feedback structures are treated efficiently. The scheduling of blocks for sequential relaxation also ignores feedback paths. Device characteristics are evaluated through a special subroutine that models a table-lookup scheme. Inside the subroutine, commonly used closed-form expressions are actually used along with artificially injected uncertainty. Both the d.c. and transient analysis routines actually use relaxation within blocks, as well as among blocks, but only over very small time windows in the case of transient analysis. The inner loop of the d.c. routine conducts a simple linear search for solutions and piecewise constant simplified models are used for evaluation in the transient routine. Extremely conservative initial bounds are used in both cases. Tolerances and time steps are chosen interactively. These algorithms are not meant to provide an efficient way to arrive at the solutions of simplified models. Since the experiments deal only with small circuits, the algorithms are designed for simple implementation.

6.3.1 Partitioning and Scheduling

The first part of the general program is a pre-processer that transforms the circuit specification into a form that can be more easily used by the bounding algorithm. The circuit must first be partitioned into strongly connected blocks, and then these blocks must be ordered for Gauss-Seidel relaxation algorithms. Since relaxation is also used within blocks, the nodes within each block must also be ordered. Once the partitioning and scheduling have been done, data structures are produced that represent the circuit in a convenient form for processing by circuit blocks. While the algorithms used here are not sophisticated, they do a reasonable job for the simple circuits considered in the experiments later.

The only input to the partitioning and scheduling subroutine is the netlist, which describes the circuit model. Each entry in the list, corresponding to a single circuit element, is denoted NET[i] for $i=0$ to p. It specifies the device type, any device parameters, and the nodes connected to the terminals. Circuit nodes must be numbered from 0 to $N-1$, where 0 corresponds to ground, 1 corresponds to VDD, and 2 through $M-1$ correspond to input voltage sources.

The output of the partitioning and scheduling subroutine is a matrix containing the node numbers with groupings and orderings appropriate for efficient circuit analysis. Each row contains the nodes that belong to one tightly coupled block of the circuit. The rows appear in the order in which the blocks should be analyzed, making sure that all strong inputs to a block are considered, when possible, before the block is considered. Within each row, the nodes within a block are similarly ordered, starting with nodes that are "closest" in terms of conductance to either ground or VDD.

In the implemented algorithm, blocks can correspond to either clusters or a pair of clusters with strong bidirectional coupling, such as would normally be found in a latch. For simplicity, the presentation here assumes that blocks correspond exactly to clusters. Adding an extra level of hierarchy to the following algorithms to account for differences is straightforward.

The first task is to group the nodes of the circuit into clusters. A convenient data structure for representing the groupings is an N element vector, called CL, where CL[i] denotes the cluster number of node i. All nodes in the same cluster have the same cluster number in their entry in the vector. The unique number used to name a particular cluster is simply the smallest node number that it contains. The vector is initialized so that CL[i]=i for $0 \leq i \leq n-1$, i.e., every node is first assumed to be a single element cluster. The netlist is then scanned element by element, in any order, looking for resistive paths between nodes. These paths can occur only through a resistor or between the drain and source of a transistor. Each time a path is found, say between nodes a and b, the two clusters that contain these nodes are merged, if they have not been already. For each combination, the new cluster is given the number of the smaller of the two original ones, guaranteeing that each will eventually have the number of its smallest node number. All mergings must occur sequentially to guarantee that the correct value of CL is produced after the netlist is exhausted. Only the independent nodes m to $n-1$ are considered when merging nodes since resistive paths through the ideal input nodes 0 to $m-1$ can not create any coupling between node behaviors.

The second task is to construct a directed graph that represents coupling between blocks. A matrix is used to represent a list of arcs in the graph. Each row corresponds to a single arc in the graph, and contains both the cluster number of its origin and of its destination. Each transistor in the circuit is scanned and the block containing its gate is connected by an arc to the block containing its source and drain, unless the two blocks are the same or have already been connected. An ordered list of blocks is initialized to the empty set. The list of unscheduled blocks is then scanned, and any whose input arcs terminate either on network input nodes, or on blocks that have already been scheduled, are added to the ordered list, i.e., are scheduled. The algorithm makes multiple passes through the list of unscheduled blocks until all are scheduled. If it gets stuck at some point, i.e., a feedback path prevents any block from being scheduled first, the first unscheduled block is forced onto the ordered block list. The nodes within each block are scheduled with a similar algorithm, where VDD and ground are treated as inputs.

6.3.2 Device Models

The simulator calculates device models with a single subroutine that returns bounds on drain current when given bounds on all transistor node voltages. The calling routine has the responsibility to scale the currents by the shape factor of the transistor. Since the transistor model has all the monotonic properties assumed in chapter four, bounds on the current are generated by evaluating the device equations with the appropriate extremes of terminal voltages.

The equations used for n-channel transistors [60] are as follows:

$$I_D = 0 \qquad \text{if } V_{GS} - V_{TE} \leq 0,$$

$$I_D = K(2V_{DS}(V_{GS} - V_{TE}) - (1 + \varphi_B)V_{DS}^2) \qquad \text{if } 0 < V_{GS} - V_{TE} < V_{DS}(1 + \varphi_B),$$

$$I_D = K(V_{GS} - V_{TE})^2/(1 + \varphi_B) \qquad \text{if } V_{GS} - V_{TE} \geq V_{DS}(1 + \varphi_B),$$

$$V_{TE} = V_{T0} + \gamma((V_{SB} + 2\varphi_{Fp})^{0.5} - (V_{BB} + 2\varphi_{Fp})^{0.5}),$$

$$\varphi_B = \gamma/(2(V_{SB} + 2\varphi_{Fp})^{0.5}).$$

The parameters are stored in a table and represent "typical" two micron NMOS and CMOS processes. The upper bound on the drain current of an n-channel transistor is calculated from an upper bound for the drain voltage, and a lower bound for the source voltage. If the upper bound on the drain voltage is greater than the lower bound on the source voltage, an upper bound on the gate voltage is used. Otherwise, a lower bound on the gate voltage is used. The program assumes that all substrates are tied to a common voltage, $-V_{BB}$, which is listed in the parameter table. The currents can be magnified by a user specified percentage to model the effect of device uncertainty. A percentage greater than 100 is only used for a positive upper bound and a negative lower bound.

6.3.3 D.C. Analysis

The d.c. analysis program uses two levels of relaxation. A solution for each block is updated by using relaxation with a partitioning down to the level of individual nodes. The initial bound of [0,VDD] is used for each node. If there is no logic feedback, each block need only be considered once to reach convergence. If there is feedback that causes multiple solutions, as in a latch, and therefore loose bounds, the final bound is displayed and the user is given the option to force tighter bounds on any node voltage. The routine is then started again using the updated bounds for an initial guess, replacing [0,VDD]. In this manner, the desired solution can be selected.

A subroutine that is used for both the d.c. and transient analysis programs calculates bounds on each node voltage derivative using the circuit model along with the appropriate extremes of other circuit variables. The results of sections 4.2 to 4.4 are used to determine the extremes. While the derivations in chapter four are quite involved, the algorithms that ultimately result are very simple. The most difficult calculation required to determine which endpoints to use in device computations is a magnitude comparison.

To update a bound on a single node voltage, the program uses the property that the current entering a node is a monotonic function of the node voltage. At the d.c. solution, the total current must be zero, so if a node voltage is found that produces a negative upper bound on current (or a negative upper bound on the voltage derivative if there is a grounded capacitor at the node), it must be a valid upper bound. A voltage step is chosen by the user and the voltage bound is shrunk in these increments to find the smallest interval that is still guaranteed to be a bound at each iteration. The program actually uses a hierarchy of grid sizes to improve on the performance of this linear search. The size of the finest voltage step determines the final accuracy of the bounds on the d.c. solution.

6.3.4 Transient Analysis

The transient analysis program uses two levels of relaxation in the same manner as the d.c. program. Relaxation within each cluster represents a simple but inefficient way to reach the solution that would be obtained using the direct methods discussed in section 4.6. The circuit transformation that duplicates each resistor and transistor inside a cluster is equivalent to ignoring the correlations among the node equations within the cluster, as is done in the inner relaxation loop. Since feedback within a cluster is positive, no additional uncertainty is produced. Since coupling is large within clusters, and more generally, within the blocks produced by the partitioning and scheduling subroutine, the relaxation within blocks is divided into small time windows specified by the user. Since the time steps are fixed, latency is not exploited.

The initial bound used is a very conservative guess for a rigorous bound. All voltages are assumed to fall in the voltage interval of $[-1,6]$ and these are used to calculate worst case voltage derivatives. The initial bound is assumed to be constant throughout the simulation time.

To calculate bounds on the behavior of a node during a time step, constant voltage bounds are first assumed for the node voltage. The method used is more sophisticated than the one presented in section 5-3, though, because each bound is treated separately. Due to the monotonic properties of the differential node equation presented in section 4.5, a lower voltage bound can be calculated by using its value to calculate its derivative. The upper bound on the voltage is not needed. The basic simplification of assuming constant node voltage can still be used if constant bounds on the trajectory of the lower bound over the time step are assumed. The assumed voltages are only allowed to vary with fixed, user-specified increments, allowing the uncertainty arising from simplified device models to be imitated. The voltage increment restriction produces an effect similar to that of using piecewise-constant device models stored in a table.

The simplification of assuming constant variables throughout each time interval makes the inner loop of the transient analysis program quite similar to the inner loop of the d.c. analysis program. If the upper bound is guaranteed to be rising or the lower bound is guaranteed to be falling, e.g., the derivative of the lower bound is negative even if it is calculated assuming the lower bound never leaves its initial value, the initial value of each bound trajectory can be used throughout the time step to calculate a bound on its slope. Otherwise, a linear search is undertaken to find the best guess for a bound on each trajectory that does not let the trajectory escape. The bounds on the derivatives of each trajectory give rise to linear bounds on the voltage waveform. These voltage bounds are then converted to constant bounds over the interval, and used to generate bounds on the voltage derivative. Now the derivative of the node voltage is considered, instead of just the derivatives of the bound trajectories.

The input waveforms are rising steps and pulses of variable width. These can be used along with inverters and other circuits to model a wide range of real inputs. Relaxation at the level of nodes is taken to convergence within a tolerance supplied at the beginning of the simulation. The relaxation at the level of blocks is under user control. Results are displayed after each iteration and the user decides if further relaxation is desired and if so, what blocks should be considered.

6.3.5 Experiments

A number of circuits were analyzed with the preceding algorithm to investigate the general behavior of bounding algorithms based on the strategy proposed in this book. The circuits were chosen to avoid situations that the simple algorithms were not designed to handle efficiently such as dynamic nodes. Even so, they illustrate many of the ideas and conjectures discussed earlier.

The first circuit considered is the chain of CMOS inverters pictured in figure 6-14. A rising step is applied to the input at t=0 ns. A time step of 0.2 ns is used along with a voltage step of 0.1 v. Figure 6-15 pictures the first three relaxations for bounds on the output voltage. The initial conditions are bounded very tightly by the d.c. algorithm with a tolerance of 0.01 v. The bounds that are produced when the Miller capacitors are removed are also pictured.

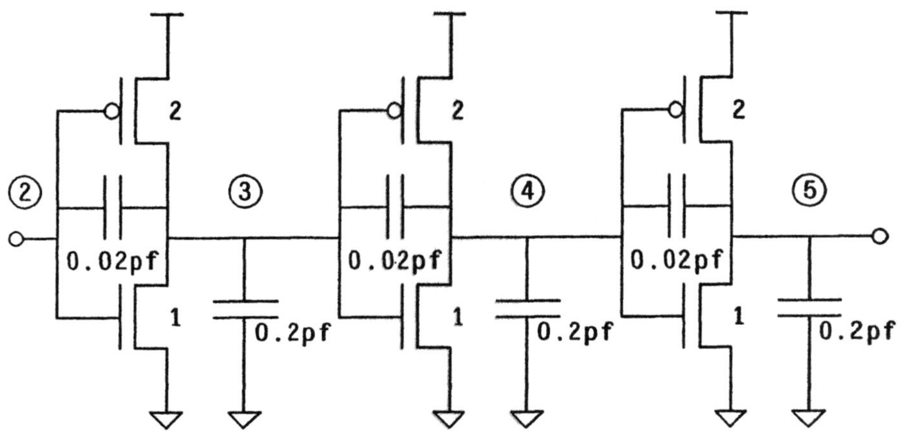

Figure 6-14: A chain of inverters with Miller capacitors.

There are two important aspects to the behavior of the bounding algorithm on the inverter chain. First, the relaxation of bounds exhibits convergence behavior similar to that of exact simulation. The third relaxation is indistinguishable from the fourth on the scale of the graph. Exact Waveform Relaxation essentially reaches convergence for restoring combinational logic in only a few passes, and the relaxing bounds have similar behavior. Relaxation can be just as efficient in bounding algorithms.

The second aspect of the results for the inverter chain is the effect of the Miller capacitance. The bounds are tighter without the small feedback between stages, but the uncertainty amplification associated with the coupling is not large. As long as the coupling is a second order effect, the uncertainty that it injects can be kept at a reasonable level. Both of the observations about the behavior of bounds on the inverter chain are valid for different levels of accuracy as well.

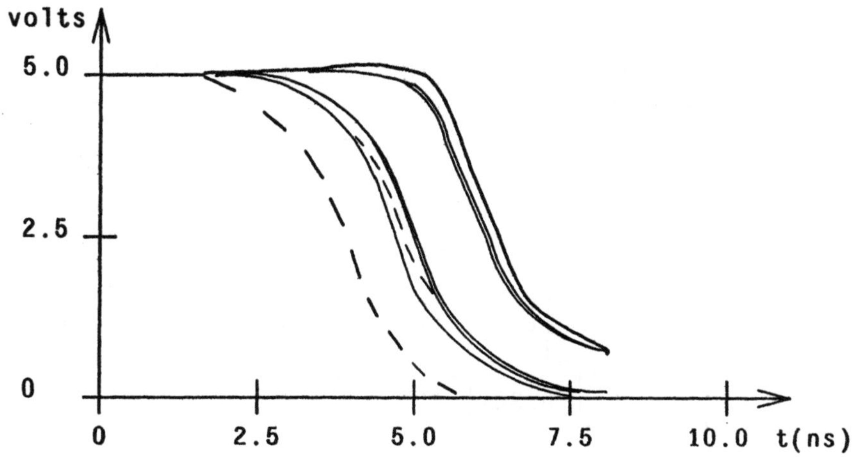

Figure 6-15: Three relaxations for an inverter chain.

The second circuit considered is the NMOS latch pictured in figure 6-16. A positive pulse is applied to the input which flips the output if it is of sufficient duration. All of the internal nodes are grouped into one block by the scheduling program as this circuit has tight feedback between two clusters.

The input is low for the d.c. solution, so without additional information the bounds on the d.c. voltages do not shrink significantly from [0,5]. The information provided to the d.c. program was that the voltage on node three was between 3 v and 5 v, and based on this, the tight bounds on the uniquely specified initial output voltage pictured in figure 6-17 are generated.

The graph shows the bounds generated for the output with a time step of 0.25 ns and a voltage step of 0.1 v, for three different input waveforms. The first input is a pulse of width 0.25 ns, and this does not flip the latch state. The second is a pulse of width 1 ns, which generates bounds on the output that diverge. The conclusion drawn from this divergence is that the second input pulse comes very close to placing the latch in its metastable state, and the bounds are not tight enough to determine which way the latch will fall after the pulse is finished. Divergence in this case is the desired answer, as it reflects uncertainty in the operation of the actual circuit that a designer would like to know. The third input is a pulse of 2 ns width that is guaranteed to flip the state of the latch.

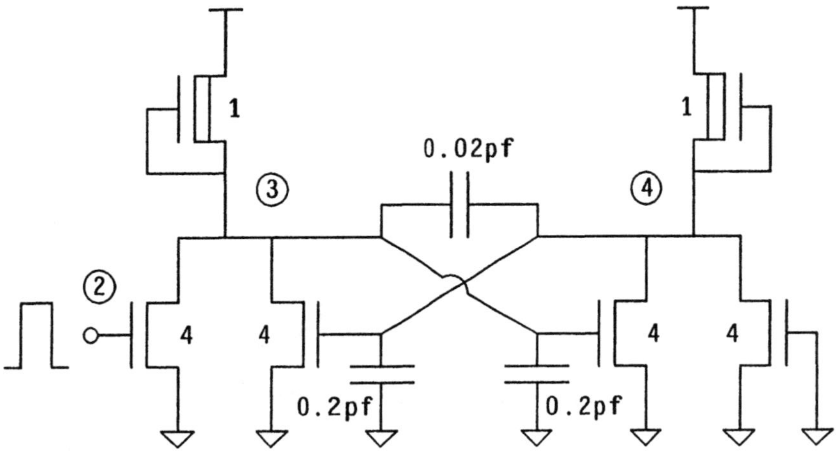

Figure 6-16: An NMOS latch.

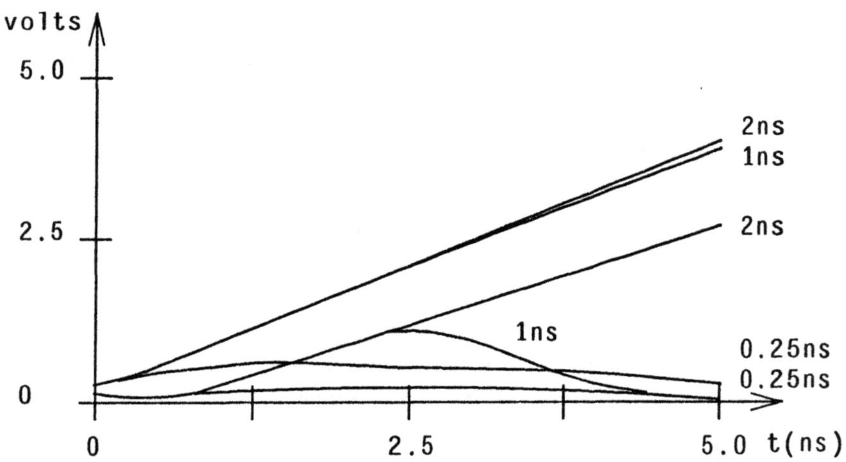

Figure 6-17: Behavior bounds for latch with three pulse widths.

The third circuit considered is the CMOS restoring logic circuit pictured in figure 6-18. The input is a rising step at t=0 ns. The lower transistor in the NAND gate pulldown is turned off fairly quickly and the output is pulled up. The pulldown transistor in the NAND gate that remains on creates a challenging situation for the simple bounding algorithm used in the transient analysis program. If the time step used for the cluster containing nodes three and four is large, the use of constant voltage bounds over the intervals keeps node three from responding quickly. The lower bound on each of the node voltages is calculated assuming the other does not move during the interval, so the bidirectional coupling between the two tends to prevent them from rising quickly.

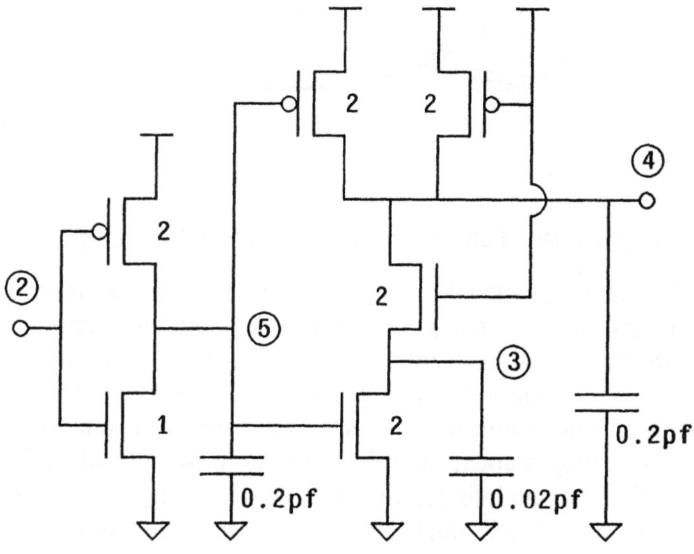

Figure 6-18: A CMOS NAND gate.

The d.c. analysis for the CMOS restoring logic circuit is straightforward. A very small voltage step is only required when gates contain long chains of transistors in the pullup or pulldown networks. The transient analysis is done with a time step of 0.05 ns and a voltage step of 0.1 v, producing the voltage bounds for each node pictured in figure 6-19. For the reasons just described, with the same level of accuracy settings, there is a large difference between the level of delay uncertainty produced for the inverter and the NAND gate.

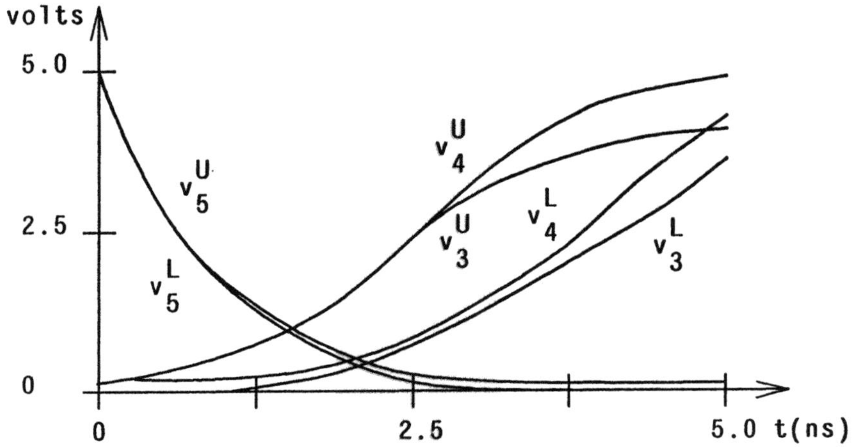

Figure 6-19: Behavior bounds for CMOS NAND gate.

The fourth circuit considered is the pass network pictured in figure 6-20. The input to the inverter is a rising step at t=0 ns and the output network is discharged thereafter. A small time step must be used for an interconnect or pass network with the simple bounding algorithm, for the same reason as in the previous circuit. The output network acts like a turned off pulldown network whenever it is being charged, and its behavior is similar when it is being discharged. The d.c. analysis requires fairly small voltage steps because the network is being pulled up through a weak depletion load pullup.

Bounds on the voltages for nodes three, five, and six are pictured in figure 6-21. A time step of 0.02 ns is used along with a voltage step of 0.02 v. Most of the delay uncertainty in the outputs comes from the uncertainty in the initial conditions caused by the pass transistors. No information is provided to the d.c. program beyond the fact that the initial input is at ground. Of course, if the outputs are charged and then discharged during a simulation, tight upper bounds will not generally rise significantly above the vicinity of four volts.

The fifth circuit considered is the three transistor XOR gate pictured in figure 6-22. This circuit is chosen because its d.c. analysis is interesting. Transient

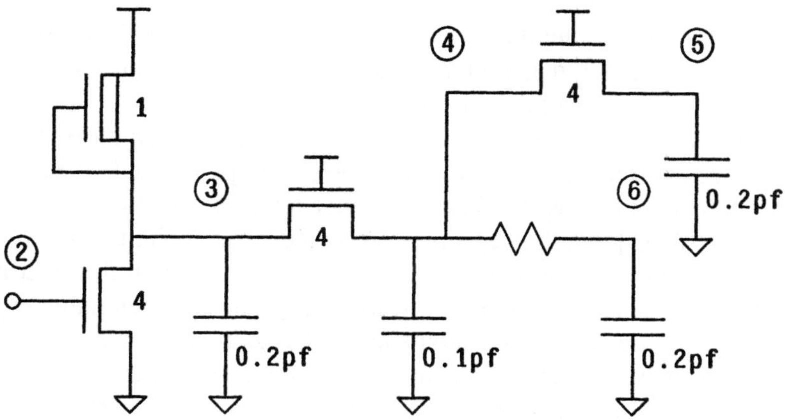

Figure 6-20: An NMOS pass network.

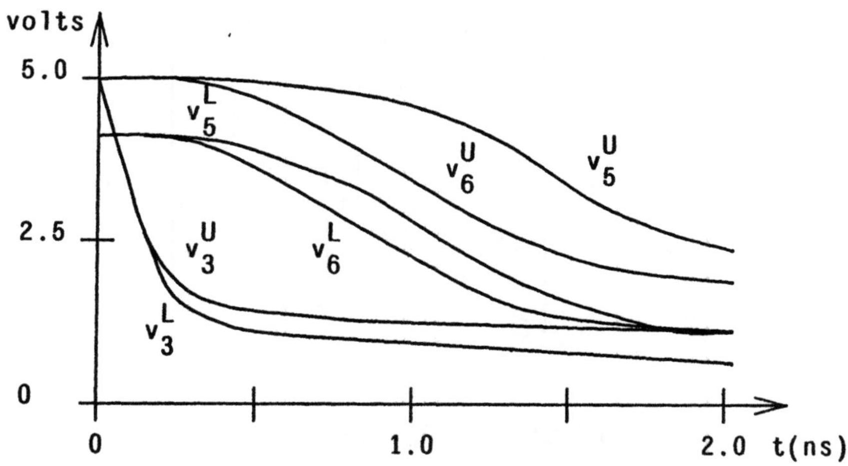

Figure 6-21: Behavior bounds for an NMOS pass network.

analysis of this circuit is not pictured because it is similar to that of more conventional NMOS combinational logic. If the input node is initially low, the two inputs to the XOR gate, nodes five and six, have opposite values. The output is therefore low, producing an inverted XOR function. This circuit is a common example of one that causes difficulty for switch-level simulators that use rough 0,X, and 1 representations for voltages. The feedback causes the internal nodes to get "stuck" in the X state.

The d.c. bounding algorithm used here can provide a rough measure of how "sticky" such a circuit is. When very large voltage steps are used, corresponding to rough representations, the bounds cannot telescope significantly from the initial interval [0,5]. As the step size is slowly reduced, a point is reached where the bounds suddenly telescope to fairly tight bounds. For this circuit, the critical voltage step size is around 0.5 v, indicating that about ten states might be required to allow a switch-level simulator to analyze this circuit in a meaningful way. This is only a rough measure as it does not account for notions of voltage "strengths" (roughly node impedances) often added to switch-level simulators.

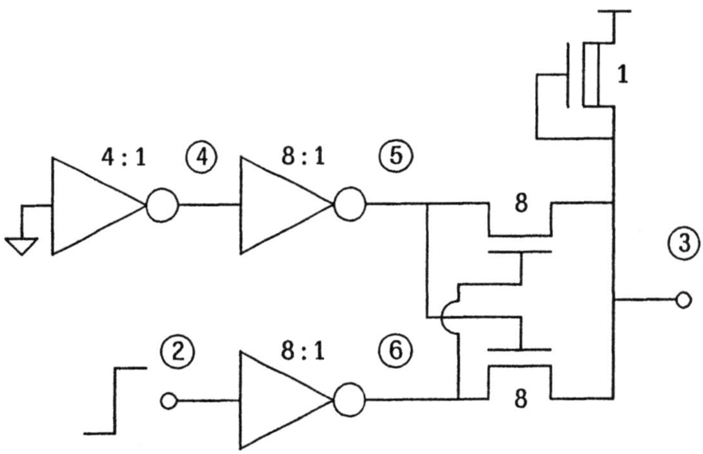

Figure 6-22: An XOR gate with feedback.

6.4 Summary

Although sophisticated bounding algorithms have not yet been developed and used on large circuits, a simple qualitative comparison between conventional approaches and bounding approaches can be made. The bounding algorithms used in the experiments conservatively model what is possible with linear or piecewise-linear models containing a similar level of accuracy, i.e., similar uncertainty caused by model simplification. In this discussion, conventional simulation is considered to be approximate, and "exact" numerical simulation is considered a special case at one end of the accuracy spectrum.

When simplified models are used to bound behaviors instead of to approximate them, roughly a factor of two is lost in computation to calculate both extremes. Additionally, the memory requirements for waveforms are at least a factor of two larger. The difference in accuracy, though, is more difficult to predict and compare.

If the delay of a single logic gate is considered, and accuracy is defined for an approximate algorithm as the standard deviation of error, accuracy can be less in the bounding approach due to the forced conservatism of the simplified models. This is independent of any uncertainty amplification that might occur due to the bounding algorithm. If the parameters of a simple model are chosen so that, over a wide range of experiments, the model's expected value of delay errors is zero, its standard deviation represents the average "roughness" of the approximation. If the parameters of the same simple model are chosen so that they are guaranteed to be conservative, one might expect the average uncertainty to be worse.

The standard deviation is not really a fair measure of accuracy for approximating simulators, though. In practice, a better definition of accuracy is enough standard deviations to instill confidence in the user. With this definition, the forced conservatism of a simplified bounding model should create comparable accuracy levels. This leaves the uncertainty amplification of the bounding strategy, along with the factor of two in computation, to compete against the statistical amplification of uncertainty found in the approximate strategy.

The conclusion that can be drawn from the experiments is that the uncertainty amplification associated with the basic bounding strategy is manageable. Isolated logic gate models can potentially be analyzed with almost no amplification, as illustrated for the one-way gate model. Also, the coupling of Miller capacitance only amplifies uncertainty by roughly a factor of two in the general algorithm.

The statistical amplification of uncertainty in approximate analysis can be quite large, as illustrated by example in section 3.4. It is possible in approximate analysis to increase accuracy in what appear to be critical paths, and obtain some ability to manage uncertainty, but the basic problem of statistical amplification will always remain.

As a result, if good uncertainty management techniques are developed for use with bounds, sophisticated bounding algorithms could potentially increase simulation efficiency for wide ranges of accuracy levels. For general digital MOS circuit simulation, piecewise-linear algorithms can be expanded to treat bounds, and the resulting bounding subroutines can be added to a more general Waveform Relaxation framework. As management algorithms improve, the bounding option could become very useful. For more specialized simulators, e.g., those tailored to consider gate arrays, special bounding algorithms are feasible.

7. CONCLUSION

As integrated circuits are growing in complexity, circuit simulation tools are being pushed towards greater flexibility and efficiency. In the long run, bounding algorithms promise to lead to sophisticated simulators that extract information about digital MOS circuit behavior with much greater efficiency than is currently possible. This book establishes a theoretical framework for generating rigorous bounding algorithms that perform well on a large portion of a typical digital MOS circuit model, i.e., the portion containing standard digital cells such as restoring logic gates. These bounding algorithms build on Waveform Relaxation, an established exact technique, and can also build on successful approximate techniques such as piecewise-linear simulation.

7.1 Future Work

The theoretical framework for bound generation presented here represents only the first step towards realizing the ultimate goal of a very efficient simulation tool. There is a great opportunity for future research projects that build upon this framework. Examples include the development of more powerful theoretical tools, of uncertainty management algorithms, of detailed bounding algorithms, and of a complete bounding simulator.

Future theoretical work in this subject falls into two main categories. The first involves extension of the circuit model to include more effects such as distributed capacitance and inductance. Distributed capacitance is a challenge due to the assumptions that were made about the use of a lumped circuit model, but it has been incorporated into bounds for linear RC circuits in the past. The incorporation of inductance is an even greater challenge, as it destroys the important monotonicity of the node equations. Extension of the circuit model to include bipolar circuits is also possible, but the performance of relaxation algorithms must be considered. The second category of theoretical work involves the derivation of more powerful results, especially in the realm of monotonic properties for general cluster circuits. There are a number of properties that, while not true in general, might be true over large time intervals. One approach to this problem is to generalize some of the results that have been obtained for the special case of RC circuits.

A bounding approach promises greater efficiency because it enables uncertainty to be managed. When running experiments with human interaction, it is easy to use partial results to guide further use of computation. Noncritical paths can be quickly eliminated and intuition about circuit operation enables sources of uncertainty to be easily located. If bounding enhancements are added to a VLSI simulation package for use in circuit design, the management of uncertainty must be automated to a large degree. A lot of interesting work in this area remains to be done as this is a complex task. At a high level, sophisticated uncertainty management algorithms might incorporate artificial intelligence techniques. At the very least, good heuristic algorithms must be developed.

The theoretical results presented in this book, along with the general bounding strategy, can lead to a wide range of detailed bounding algorithms. The simple algorithms used for the experiments presented here represent only the very beginning of an interesting search for efficient detailed strategies. An example of a more ambitious exploration in this area would be to combine algorithms from a Waveform Relaxation simulator and a piecewise-linear simulator in a bounding context. Improvements such as adding bounds on charge sharing to improve performance for dynamic circuits would also be extremely useful. The primary goal of any bounding algorithm is efficiency, as it is motivated by a desire to trade accuracy for more speed. As such, work is also needed at the lowest levels of bounding algorithms to optimize efficiency, including the use of table lookup schemes and pre-compilation of circuits.

An important project that would help direct other future research is the development of an experimental bounding simulator. The simple algorithms developed in this book are already powerful enough to be directly useful in many cases, especially in gate array simulation. A simulator must be used by designers to gain experience and to guide work on improvements. By making the simulator modular and flexible, one could use it to facilitate experiments with new algorithms. As a result, such a project could be useful in conjunction with any of the other future research projects already mentioned.

7.2 Closing Comments

More efficient approaches to VLSI circuit simulation can significantly reduce the cost of designing large integrated circuits. A bounding approach has many advantages, and initial experiments regarding feasibility have been very encouraging. Although a large amount of work remains to be done to fully exploit the bounding approach, this future work promises a significant contribution to the field of VLSI circuit simulation.

References

[1] L. W. Nagel, "SPICE2: A Computer Program to Simulate Semiconductor Circuits," ERL Memo ERL-M520, Univsity of California, Berkeley, May 1975.

[2] William T. Weeks, Alberto J. Jimenez, Gerald W. Mahoney, Deepak Mehta, Hassan Qassemzadeh, and Terence R. Scott, "Algorithms for ASTAP- A Network Analysis Program," *IEEE Transactions on Circuits Theory,* Vol. 11, 1973, pp. 628-634.

[3] Albert E. Ruehli and Gary S. Ditlow, "Circuit Analysis, Logic Simulation, and Design Verification for VLSI," *Proceedings of the IEEE,* Vol. 71, No. 1, January 1983, pp. 34-48.

[4] B. Chawla, H. Gummel, and P. Kozak, "MOTIS- An MOS Timing Simulator," *IEEE Transactions on Circuits and Systems,* Vol. 12, 1975, pp. 301-310.

[5] S. P. Fan, M. Y. Hsueh, A. R. Newton, and D. O. Pederson, "MOTIS-C: A New Simulator for MOS LSI Circuits," *Proceedings of the IEEE International Symposium on Circuits and Systems,* IEEE, April 1977, pp. 700-703.

[6] A. Richard Newton, "Techniques for the Simulation of Large-scale Integrated Circuits," *IEEE Transactions on Circuits and Systems,* Vol. CAS-26, No. 4, April 1979, pp. 741-749.

[7] Resve A. Saleh, James E. Kleckner, and A. Richard Newton, "Iterated Timing Analysis in SPLICE1," *Proceedings of IEEE International Conference on Computer-Aided Design,* IEEE, September 1983, pp. 139-140.

[8] Guido Arnout and Hugo J. DeMan, "The Use of Threshold Functions and Boolean-Controlled Network Elements for Macromodeling of LSI Circuits," *IEEE Journal of Solid State Circuits,* Vol. SC-13, No. 3, June 1978, pp. 326-332.

[9] P. H. Reyneart, H. DeMan, G. Arnout, and J. Cornelissen, "DIANA, A Mixed-mode Simulator with a Hardware Description Language for Hierarchical Design of VLSI," *Proceedings of the IEEE International Conference on Circuits and Computers*, IEEE, October 1981, pp. 356-360.

[10] N. B. Guy Rabbat, Alberto Sangiovanni-Vincentelli, and Hsueh Y. Hsieh, "A Multilevel Newton Algorithm with Macromodeling and Latency for the Analysis of Large-Scale Nonlinear Circuits in the Time Domain," *IEEE Transactions on Circuits and Systems*, Vol. CAS-26, No. 9, September 1979, pp. 733-740.

[11] Ekachai Lelarasmee, Albert Ruehli, and Alberto Sangiovanni-Vincentelli, "The Waveform Relaxation Method for Time-Domain Analysis of Large Scale Integrated Circuits," *IEEE Transactions on Computer-Aided Design of Integrated Circuits and Systems*, Vol. 3, 1982, pp. 131-145.

[12] Ekachai Lelarasmee and Alberto Sangiovanni-Vincentelli, "RELAX: A New Circuit Simulator for Large Scale MOS Integrated Circuits," *Nineteenth Design Automation Conference Proceedings*, ACM-IEEE, June 1982, pp. 682-690.

[13] Jacob White and Alberto Sangiovanni-Vincentelli, "RELAX2: A Modified Waveform Relaxation Approach to the Simulation of MOS Digital Circuits," *Proceedings of IEEE International Symposium on Circuits and Systems*, IEEE, May 1983, pp. 756-759.

[14] Randal E. Bryant, "An Algorithm for MOS Logic Simulation," *Lambda*, Vol. 4, 1980, pp. 22-30.

[15] Randal E. Bryant, "A Switch-Level Model and Simulator for MOS Digital Systems," *IEEE Transactions on Computers*, Vol. C-33, No. 2, February 1984, pp. 160-177.

[16] Randal Bryant, *A Switch Level Simulation Model for Integrated Logic Circuits*, PhD dissertation, MIT, 1981.

[17] Christopher J. Terman, "RSIM - A Logic-Level Timing Simulator," *Proceedings of IEEE International Conference on Computer Design: VLSI in Computers*, IEEE, October 1983, pp. 437-440.

[18] Christopher J. Terman, *Simulation Tools for Digital LSI Design,* PhD dissertation, MIT, 1983.

[19] Carver Mead and Lynn Conway, *Introduction to VLSI Systems,* Addison-Wesley Publishing Company, Inc., Reading, Mass., 1980.

[20] J. Ousterhout, "Crystal: A Timing Analyzer for nMOS VLSI Circuits," *Third CALTECH Conference on VLSI,* Computer Science Press, March 1983, pp. 57-70.

[21] N. Jouppi, "TV: An nMOS Timing Analyzer," *Third CALTECH Conference on VLSI,* Computer Science Press, March 1983, pp. 71-86.

[22] Mark D. Matson, "Macromodeling of Digital MOS VLSI Circuits," *Twenty-Second Design Automation Conference Proceedings,* ACM-IEEE, June 1985, pp. 144-151.

[23] Willem M. G. van Bokhoven, *Piecewise-Linear Modelling and Analysis,* Kluwer Technische Boeken, 1981.

[24] J. T. J. van Eijndhoven, *A Piecewise Linear Simulator for Large Scale Integrated Circuits,* PhD dissertation, Eindhoven University of Technology, 1984.

[25] Willem M. G. van Bokhoven, "A Transparent Level CMOS Simulator," *Proceedings of IEEE International Conference on Computer Design: VLSI in Computers,* IEEE, October 1984, pp. 37-41.

[26] V. B. Rao, T. N. Trick, and I. N. Hajj, "A Table-Driven Delay-Operator Approach to Timing Simulation of MOS VLSI Circuits ," *Proceedings of IEEE International Conference on Computer Design: VLSI in Computers,* IEEE, October 1983, pp. 445-448.

[27] M. E. Zaghloul and P. R. Bryant, "Error Bounds of Solutions of Nonlinear Networks When Using Approximate Element Characteristics," *IEEE Transactions on Circuit and Systems,* Vol. CAS-27, January 1980, pp. 20-29.

[28] Charles A. Desoer and H. Haneda, "The Measure of a Matrix as a Tool to Analyze Computer Algorithms for Circuit Analysis," *IEEE Transactions on Circuit Theory,* Vol. CT-19, September 1972, pp. 480-486.

[29] I. W. Sandberg, "Some Theorems on the Dynamic Response of Nonlinear Transistor Networks," *The Bell System Technical Journal,* Vol. 48, No. 1, January 1969, pp. 35-54.

[30] Jorge Rubinstein, Paul Penfield, and Mark Horowitz, "Signal Delays in RC Tree Networks," *IEEE Transactions on Computer Aided Design,* Vol. CAD-2, No. 3, July 1983, pp. 202-211.

[31] John L. Wyatt, Jr., Qingjian Yu, Charles A. Zukowski, Han-Ngee Tan, and Peter O'Brien, "Improved Bounds on Signal Delay in MOS Interconnect," *Proceedings of IEEE International Symposium on Circuits and Systems,* IEEE, May 1985, pp. 903-906.

[32] John L. Wyatt, Jr. and Qingjian Yu, "Signal Delay in RC Meshes, Trees and Lines," *Proceedings of 1984 International Conference on Computer-Aided Design,* IEEE, November 1984, pp. 15-17.

[33] John L. Wyatt, Jr., Charles A. Zukowski, and Paul Penfield, Jr., "Step Response Bounds for Systems Described by M-Matrices, with Application to Timing Analysis of Digital MOS Circuits," *Proceedings of 24th Conference on Decision and Control,* IEEE, December 1985, pp. 1552-1557.

[34] John L. Wyatt, Jr., "Monotone Sensitivity of Nonlinear Nonuniform RC Transmission Lines, with Application to Timing Analysis of Digital MOS Integrated Circuits," *IEEE Transactions on Circuits and Systems,* Vol. CAS-32, No. 1, January 1985, pp. 28-33.

[35] Qingjian Yu and Omar Wing, "Waveform Bounds of Nonlinear RC Trees," *Proceedings of IEEE International Symposium on Circuits and Systems,* IEEE, May 1984, pp. 356-359.

[36] Mark Horowitz, "Timing Models for MOS Pass Networks," *Proceedings of International Symposium on Circuits and Systems,* May 1983, pp. 198-201.

[37] Lance A. Glasser, "The Analog Behavior of Digital Integrated Circuits," *Eighteenth Design Automation Conference Proceedings,* ACM-IEEE, July 1981, pp. 603-612.

[38] Charles A. Zukowski, *The Bounding Approach to VLSI Circuit Simulation,* Kluwer, 1986.

[39] E. R. Hansen, "A Generalized Interval Arithmetic," *Proceedings of the International Symposium on Interval Mathematics,* 1975, pp. 7-18.

[40] John Paulos, *Measurement and Modeling of Small-Geometry MOS Transistor Capacitances,* PhD dissertation, MIT, August 1984.

[41] C. A. Desoer and J. Katzenelson, "Nonlinear RLC Networks," *Bell System Technical Journal,* Vol. 44, 1965, pp. 161-198.

[42] Paul Penfield, Jr., Robert Spence, and S. Duinker, *Tellegens Theorem and Electrical Networks,* MIT Press, 1970, p. 42.

[43] Louis Weinberg, *Network Analysis and Synthesis,* McGraw-Hill, Inc., 1962, p. 336.

[44] Mituhiko Araki, "M-Matrices (Matrices with Nonpositive Off-diagonal Elements and Positive Principle Minors)," Publication 74/19, Dept. of Comp. and Control, Imperial College, London, March 1974.

[45] John L. Wyatt, Jr., "Signal Delay in RC Mesh Networks," *IEEE Transactions on Circuits and Systems,* Vol. CAS-32, No. 6, June 1985.

[46] R. M. Cohn, "The Resistance of an Electrical Network," *Proceedings of the American Math Society,* AMS, June 1950, pp. 316-324.

[47] Charles A. Zukowski and John L. Wyatt, Jr., "Sensitivity of Nonlinear One-Port Resistor Networks," *IEEE Transactions on Circuits and Systems,* Vol. CAS-31, No. 12, December 1984, pp. 1048-1051.

[48] J. White, A. S. Vincentelli, F. Odeh, A. Ruehli, "Waveform Relaxation: Theory and Practice," *Transactions of The Society for Computer Simulation,* Vol. 2, No. 2, June 1985, pp. 95-134.

[49] Wilfred Kaplan, *Ordinary Differential Equations,* Addison-Wesley Publishing Company, Inc., 1958, p. 477.

[50] Walter Rudin, *Principles of Mathematical Analysis,* McGraw-Hill, Inc., 1976, p. 220.

[51] Charles A. Desoer and Ernest S. Kuh, *Basic Circuit Theory,* McGraw-Hill, Inc., 1969.

[52] I. W. Sandberg, "A Nonnegativity-Preservation Property Associated with Certain Systems of Nonlinear Differential Equations," *Proceedings 1974 IEEE International Conference on Systems, Man and Cybernetics,* IEEE, 1974, pp. 230-233.

[53] John L. Wyatt, Jr., "Monotone Behavior of Nonlinear RC Meshes," VLSI Memo 83-128, MIT, November 1982.

[54] John L. Wyatt, Jr. and Paul Bassett, "Spatial Monotonicity and Positive Invariance for Nonlinear RC Lines and Trees," VLSI Memo 83-140, MIT, April 1983.

[55] John L. Wyatt, Jr., "Partial Ordering and Monotone Sensitivity for Nonlinear RC Meshes and Lines," VLSI Memo 83-145, MIT, June 1983.

[56] R. E. Moore, *Interval Analysis,* Prentice-Hall, 1966.

[57] Charles Zukowski, "Relaxing Bounds for Linear RC Mesh Circuits", to appear in IEEE Transactions on Computer-Aided Design.

[58] T.-M. Lin and C. Mead, "Signal Delay in General RC Networks with Application to Timing Simulation of Digital Integrated Circuits," *Proceedings, Conference on Advanced Research in VLSI,* MIT, January 1984, pp. 93-99.

[59] R. John Kaye and Alberto Sangiovanni-Vincentelli, "Solution of Piecewise-Linear Ordinary Differential Equations Using Waveform Relaxation and Laplace Transforms," *IEEE Transactions on Circuits and Systems,* Vol. CAS-30, No. 6, June 1983, pp. 353-357.

[60] L. Glasser and D. Dobberpuhl, *The Design and Analysis of VLSI Circuits,* Addison-Wesley Publishing Company, Inc., 1985.

INDEX

Approximating simulators, 10, 175
ASTAP, 5

Bipolar, 210
Bootstrap drivers, 49
Bound generation, 30
Bound performance, 41
Bound relaxation, 35, 139
Bound specification, 16
Bounding applications, 50
Bounding approaches, 12, 126
Bounding function, 143
Bounding simplification, 18, 41

Capacitor model, 65, 97
Charge sharing, 24, 162, 187, 192, 210
Circuit behaviors, 26, 141
Circuit excitations, 21
Circuit models, 25, 59
Circuit simulation, 1, 5, 29
Cluster circuits, 125, 162
Conservative generation, 18
Conservative specification, 18
Critical paths, 23, 52

Digital MOS simulators, 7
Diode configuration, 63, 84, 90, 92
D. C. solution, 23, 44, 139, 149, 197
Distributed capacitance, 64, 66, 209
Dynamic storage, 24, 43, 199, 210

Eigenvalues, 193
Element sensitivity, 69

Fabrication process variations, 3, 53
Fault simulation, 51
Feedback, 3, 35, 41, 134, 150, 194
Fixed point, 141, 149, 166

Hysteresis, 47

Inclusion monotonic, 143, 152
Inductance, 21, 71, 209
Input-independent analysis, 55
Interval arithmetic, 3
Introverted intervals, 146, 152, 158
Iteration function, 141

Latches, 9, 44, 139, 150, 201
Latency, 1, 7, 198
Linear models, 2, 3, 28, 126, 186, 207
Local connections, 100
Local time constants, 164, 187
Logic simulation, 10, 29

Macromodels, 12, 137
Memory, 192
M-matrices, 72
Miller capacitance, 3, 8, 24, 35, 126, 155, 174, 200
Mixed-mode simulation, 8, 51
Model relaxation, 122
MOSSIM, 11, 29
Murphy's law, 15

Network topology, 66
Nodal analysis, 71, 98, 116
Node circuits, 117
Nonreciprocal capacitors, 66
Numerical roundoff, 31

One-port grouping, 75, 84

Partial ordering, 18
Partitioning, 1, 7, 111, 116, 137, 194
Pass transistors, 13, 43, 192, 204
Picard iteration, 101, 158
Piecewise-linear models, 2, 3, 28, 207
Power supply, 21

RC meshes, 13, 140, 161, 186
RC trees, 2, 12, 126, 130, 132
RELAX, 8
Relaxation, 41, 125, 139, 187, 194
Resistor model, 60, 70
Restoring logic, 41, 130, 184, 200, 203
Ring oscillators, 9, 32, 46
RSIM, 11

Scheduling, 194
Schmitt trigger, 47, 91
Sense-amp, 192
Series-parallel group, 80, 87, 110
Shape factor, 174
SHRINK, 189
Sparse matrices, 100
Speed-accuracy tradeoff, 10, 32
SPICE, 5, 174, 185
SPLICE, 8
Statistical accuracy, 52
Substrate terminal, 62, 84, 86, 90, 94
Switch-level simulation, 10, 29, 51, 206

Table-lookup, 7
Terminal constraints, 68
Time windows, 9, 44, 46, 48, 194
Timing verification, 2, 51, 55
Timing analysis, 29, 50
Transient analysis, 154, 198
Transistor clusters, 9, 35, 41, 137, 195, 210
Transistor model, 62, 84, 197

Uncertainty management, 50, 208, 210

Voltage minimax theorem, 72, 106, 109

Waveform Relaxation, 4, 8, 39, 59, 100, 116, 209
Worst case analysis, 53

MIX
Papier aus verantwortungsvollen Quellen
Paper from responsible sources
FSC® C105338
FSC
www.fsc.org

If you have any concerns about our products,
you can contact us on
ProductSafety@springernature.com

In case Publisher is established outside the EU,
the EU authorized representative is:
Springer Nature Customer Service Center GmbH
Europaplatz 3, 69115 Heidelberg, Germany

Printed by Libri Plureos GmbH
in Hamburg, Germany